Astronomers' Observing Guides

Series Editor
Dr. Michael D. Inglis, BSc, MSc, Ph.D.
Fellow of the Royal Astronomical Society
Suffolk County Community College
New York, USA
inglism@sunysuffolk.edu

For further volumes:
http://www.springer.com/series/5338

Peter Grego

Mars and How
to Observe It

 Springer

Peter Grego
Cornwall
PL26 8AS
United Kingdom

ISSN 1611-7360
ISBN 978-1-4614-2301-0   ISBN 978-1-4614-2302-7 (eBook)
DOI 10.1007/978-1-4614-2302-7
Springer New York Heidelberg Dordrecht London

Library of Congress Control Number: 2012935783

Printed on acid-free paper

Springer is part of Springer Science+Business Media (www.springer.com)

To Auntie Pat, a rock firmer than Mars itself

Introduction: A Perspective on Mars

Discover Mars

Mars was discovered way back in the spring of 1982 – April 10th at 10.15 pm, to be precise. Following a pleasant evening's observations of Jupiter and Saturn, and looking forward to a later view of the rising waning gibbous Moon, a young amateur astronomer delighting in the thrills afforded him by his modest 60 mm refractor, swung the little instrument towards a bright orange star, high in the south. Being relatively new to the heavens, the novice stargazer did not suspect the true nature of this amber-tinted luminary which shone in western Virgo, trailing the feet of mighty Leo.

Using the highest magnification available – a giddying power of 100× – the gleaming orange fleck was brought into focus. Contrary to expectations, this was no star; instead, a tiny disk presented itself, a marvelous ruddy circle upon whose face was arrayed a set of distinct dusky markings. Instantly recognizable as the planet Mars, the young astronomer excitedly fumbled with his pencil and sketchpad while at the same time juggling his red torch, and strained his eye at the telescope eyepiece to discern as much detail as possible and set it down in an observational drawing.

Admittedly, keeping the planet within the somewhat restricted field of view presented by the Huygenian eyepiece by frequent tugs on the tube of un-driven telescope tube presented almost as much of a challenge as the observational drawing itself; but experience had equipped him with the skills necessary to track celestial objects using this shakiest of altazimuth-mounted instruments. Closer examination revealed a broad V-shaped marking at the center of the planet, accompanied by less prominent shadings to its sides and spreading to the north, where a brilliant white spot crowned the disk. My first sighting of Mars closely matches Christiaan Huygens' observation of 28 November 1659, the first known telescopic observation of the planet to show its features. That large 'V-shaped' marking was none other than Syrtis Major, the planet's most prominent feature; the white spot, its north polar ice cap.

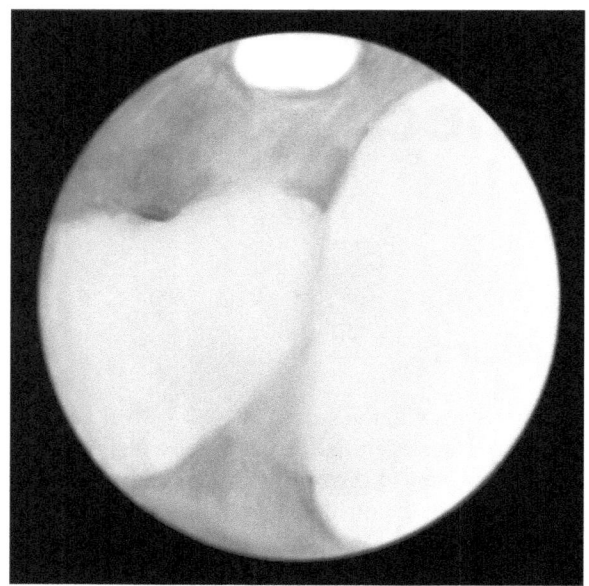

The author's first observation of Mars, made on 10 April 1982 at 21:15 UT, with Syrtis Major approaching the CM. CM 279°. P 30°. Tilt 23.3°. Phase 99%. Diameter 14.7″. Magnitude −1.3. Observing notes: A most remarkable sight! My first vision of the Red Planet through a telescope. The cap was underlain by a shady, dusky cusp, the western side being the most markedly defined. The markings at the south certainly seemed to be a dusky finger-like obtrusion (Credit: Grego)

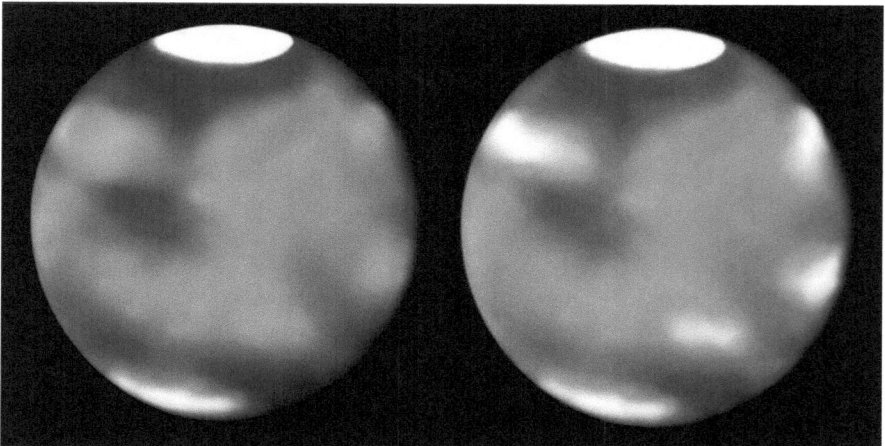

2010 January 13, 23:30 UT. CM 171°. P 1°. Tilt 16.8°. Phase 99%. Diameter 13.7Đ. Magnitude −1.1. In both integrated light and in yellow and blue filters, the large north polar cap is brilliant with a dusky border extending south through Phlegra to a dusky but poorly defined Trivium Charontis. In blue light several other bright areas stand out prominently — one in the Elysium area to the northwest of Trivium Charontis, another in the Memnonia region bordering Mare Sirenum, one in the southern Tharsis region near the terminator and another near the terminator but further north in the vicinity of Ascraeus Lacus. The Electris region along the southern limb was also bright. 200 mm SCT, 250×, integrated light and yellow W12 (*left*) and blue W80A (*right*) (Credit: Grego)

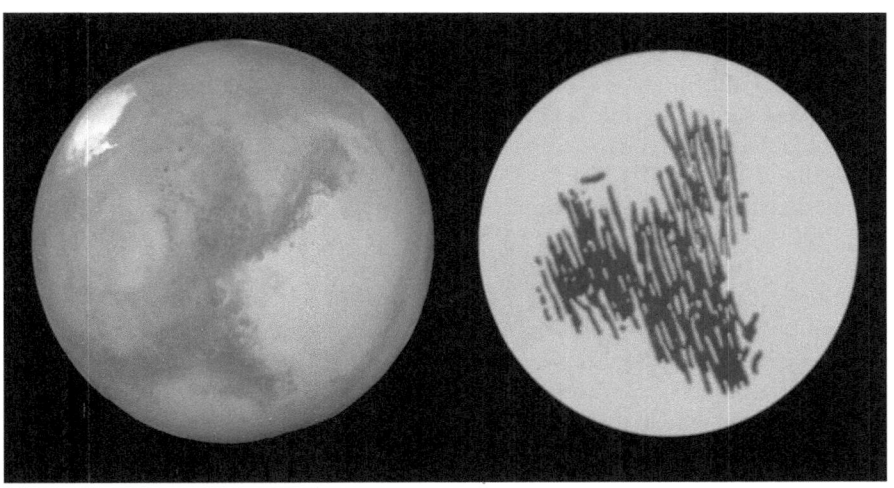

Hubble Space Telescope image of Mars taken on 26 August 2003 compared with Huygens' observational drawing of Mars, made on 28 November 1659 (Credit: NASA/public domain)

Anyone who has ever put their eye to the telescope and viewed the night skies has experienced their own special thrills of astronomical discovery, and few celestial objects are as thrilling to discover as the Mars, the Red Planet, a world of perennial mystery and the inspiration of boundless human speculation and fantasy through the centuries.

Since my own personal 'discovery' of the Red Planet in 1982, described above, I've eagerly followed each apparition – all 14 of them – through the eyepiece of a variety of telescopes, including the memorable apparition of 2003–2004 when Mars swung in to approach the Earth more closely than it had done for 50,000 years. Each apparition has shown a slightly different Mars, with variations in the apparent angular size of the planet at opposition, its position in the heavens, its axial tilt and phase; each apparition brings with it actual differences in the intensity and shape of Mars' surface markings, along with transient atmospheric phenomena including dust storms and cloud features.

Mars is by no means worthy of the epithet once attributed to it by one famous astronomer, who wrote that, observationally, Mars is the biggest disappointment in the Solar System. On the contrary – Mars is an utterly intriguing world with more than enough visual appeal to hold the observer's attention. While its broad features are discernable through small amateur telescopes, its finer features present themselves only to those who give it the time and attention that it deserves.

A World in Dichotomy

In many respects, both literally and figuratively, Mars is a planet of two halves. Our scientific knowledge of the planet can be placed into two broad categories, the line of demarcation beginning in 1971 when NASA's Mariner 9 first imaged the planet from up close in 1971. Before this epoch-making technological achievement, Mars had really only been known intimately through telescopic observations, beginning in 1636 with Francisco Fontana when he made the first disc drawing of Mars.

To this day the amateur observer still sees Mars as it was known to the eyes of the great observers of the past – observers such as Christiaan Huygens, Gian Domenico Cassini, William Herschel, Giovanni Schiaparelli, Edward Emerson Barnard and Eugene Antoniadi, to name but a few. Indeed, visual observers still use the Martian system of nomenclature established by Schiaparelli in the late nineteenth century; his aim was to produce an unbiased form of nomenclature was based upon ancient Latin and Mediterranean place names, Biblical and other mythological sources. He wrote: 'These names may be regarded as a mere artifice… After all, we speak in a similar way of the seas of the Moon, knowing very well that they do not consist of liquid masses.'

This rich nomenclature, however, was of course given only to telescopically visible features, the albedo features formed from light and dark areas. When Mars is at its largest each apparition, around the time of opposition, the telescopic observer is seeing a planetary disc which is fully illuminated; features of relief aren't visible, as there are no shadows thrown up by the landscape. Once space probes began to explore Mars, they imaged far more than albedo features – the old nomenclature was simply unable to be applied to the new Mars, with its freshly-discovered craters, faults, valleys, hills, mountains and other landscape features. A standard nomenclature based on types of terrain and feature was agreed by the IAU (International Astronomical Union), and there was retained a good deal of congruence with the old nomenclature. For example, the large dark V-shaped feature that visual observers know as Syrtis Major (named after the Mediterranean Gulf of Sidra) was initially renamed Syrtis Major Planitia (Syrtis Major plain) but then, following the discovery that it was actually a large shield volcano of low relief, renamed Syrtis Major Planum (Syrtis Major plateau).

Visual observers had long known that the globe of Mars is split between north and south; extensive bright areas, bound by subtle shadings, spread across the planet's southern hemisphere, while the majority of well-defined dark features dominate the northern hemisphere. Under closer examination by space probes, the north–south dichotomy exists in terms of topographical features and geology. Mars' southern hemisphere is covered with thousands of sizeable impact craters, while the north is relatively crater-free; many ancient craters in the north have been completely obliterated by smooth lava plains. On average, the southern hemisphere is 3 km higher than the northern hemisphere.

In order to be consistent with the system of longitude used by visual observers, and for ease of reference, all longitude co-ordinates in this book are given west of the Martian prime meridian.

In terms of human arts, fantasy and imagination, the Martian split reflects the dichotomy between pre-Space Age Mars and the planet as known to us today. In the past, our relative lack of knowledge about the conditions existing on Mars – the nature of its surface and its atmosphere, the seeming possibility that advanced forms of life might exist there – gave writers virtual *carte-blanche* to create the most elaborate fiction involving Martian civilizations. Generally supposed to be an older planet than the Earth, sometimes looked upon as a planet in physical decline, some of these civilizations were benign, others intent on war, both civil and interplanetary. Perhaps the most famous expression of a belligerent Martian civilization was created by H.G. Wells in his book *The War of the Worlds* (1898) which saw the attempt by a technologically advanced race of Martians to conquer the Earth and destroy all humans. Ultimately, these 'cool and unsympathetic' aliens were beaten by the lowliest of our planet's life forms – bacteria. In a strange twist of history, it now seems possible that life on Mars, if it exists at all, is in the form of bacteria and other simple life.

Speculation was not restricted to writers of fiction; many astronomers imagined that it was probable that some sort of life existed on Mars. Indeed, observed seasonal changes on Mars appeared to suggest this; we can still observe these seasonal changes, but we now know their cause to be other than life. In the most extreme case, that of the wealthy amateur Percival Lowell, it was speculated that an advanced Martian civilization had created a vast network of interconnected canals to channel melted polar waters in order to irrigate the planet's dry plains. Lowell elaborated on his theories in his books *Mars* (1895), *Mars and its Canals* (1906) and *Mars as the Abode of Life* (1908).

Undeterred by the facts as later revealed by a host of probes – orbiters, landers and rovers – the Mars of the Space Age was (and is being) used as the setting for modern fictional excursions into fantasy. One of the most notable of these, the action-adventure movie *Total Recall* (1990), sees the discovery of a now-extinct Mars civilization whose technology is finally used to convert the hostile Red Planet into a world with comfortable climes and a breathable atmosphere. A long-dead Martian civilization is also the subject of a variety of 'conspiracy theories' which claim that space probe images have revealed extraordinary structures built by the Martians. Most famous of them is the so-called 'Face of Mars', a hill in Cydonia, shown in an image by Viking 1 as resembling an odd sort of face; higher resolution images of the feature demolish any notion that the hill is anything but a natural formation.

Mars, and How to Observe It

Tempting as it is to delve deeply into the rich history of Mars observation, the remit of this series is to present in the first part of the book what is currently known about any particular celestial object or phenomenon, and then in the second part to offer guidance on how best to go about making observations of it. I'm gratified that my works in this series – *The Moon and How to Observe It* (Springer, 2005) and *Venus and Mercury and How to Observe Them* (Springer, 2008) – have been well-received, yet some reviewers appear to have missed this most essential point about this series.

In writing this book, it's been extremely challenging to avoid fully recognizing the hundreds of careful observers of Mars, many of them great visual observers whose contributions to astronomy were been immense. Those observers, some of whom are cited above, have been difficult to avoid, ingrained as they are into the fabric of the history of the planet's observation. Not only has it been a tough call to neglect to liberally mention the accomplishments and insights of these figures throughout the text, it hasn't been possible to present many of their observational drawings, even though their work might be considered to be of direct relevance to visual work today; indeed, their observations remain a topic of discussion by amateur astronomers the world over. To have referred extensively to historical observations would have required a much larger book, and one at variance with the aims of this series. However, should the reader be interested in finding out more about historical observations of Mars (a pursuit to be thoroughly recommended) I have made a few choice selections for further reading at the end of this book.

Similarly, it has been necessary to omit intricate details about the many space probes that have orbited Mars and investigated its satellites, soft-landed on the Martian surface and roved around hill, dale and crater, imaging, sniffing, scratching, drilling and sampling here and there. Great use has, however, been made of the important data and the highly revealing images obtained by these probes, keeping it as up-to-date as possible with the inclusion of some of the very latest in-situ Mars images and data. Again, I have recommended a number of books for further reading which I think will enlighten the reader on the incredible history of the robotic exploration of the Red Planet.

Inspiration from the Red Planet

With all this in mind, this book intends to shed light upon Mars' physical nature and its phenomena, while the means and techniques to make meaningful observations are explained clearly. Importantly, I have attempted to bridge the gap between the physical Mars, as revealed in glorious detail by spaceprobes, and that quite different looking world that hovers in the telescope eyepiece, teasing us with its dusky markings and challenging us to capture its subtleties. I sincerely hope that this modest contribution to the existing literature on Mars inspires a new generation of observers to take to the eyepiece and thereby gain a first-hand view of the most Earth-like planet that we know of.

There can be few more inspirational words written about observing Mars than the following two of my own favorite Mars quotations, both written more than a century ago, one by an astronomer, the other, an author, appropriately enough:

Almost as soon as magnification gives Mars a disk, that disk shows markings, white spots crowning a globe spread with blue-green patches on an orange ground.

Percival Lowell

I still remember that vigil very distinctly: the black and silent observatory, the shadowed lantern throwing a feeble glow upon the floor in the corner, the steady ticking of the clockwork of the telescope, the little slit in the roof – an oblong profundity with the stardust streaked across it. Ogilvy moved about, invisible but audible. Looking through the telescope, one saw a circle of deep blue and the little round planet swimming in the field. It seemed such a little thing, so bright and small and still, faintly marked with transverse stripes, and slightly flattened from the perfect round. But so little it was, so silvery warm – a pin's-head of light! It was as if it quivered, but really this was the telescope vibrating with the activity of the clockwork that kept the planet in view.'

From The War of the Worlds, by H.G. Wells

A planet that captures the imagination. The modern observer is unlikely to have a view of Mars like this one, based upon a Hubble Space Telescope image overlain with observations by Percival Lowell. A century ago, some observers placed great faith in the reality of the planet's 'canals' – features we now know to be largely illusory (Credit: Peter Grego)

Peter Grego
St Dennis, Cornwall, UK
August 2011

About the Author

Peter Grego is an astronomy writer and editor. A regular watcher of the night skies since 1976, he observes from his home in St. Dennis, Cornwall, UK with a variety of instruments. His telescopes include a 102 mm refractor, home-made 150 and 300 mm Newtonians (telescope mirror-making is another of his interests) and a 445 mm Newtonian, but his most-used instrument is his 200 mm SCT. Grego's primary observing interests are the Moon's topography, Mars and Jupiter, but he likes to 'go deep' when there's no lunar glare to contend with. He now likes to use a hand-held computer to make observational drawings.

Grego has directed the Lunar Section of Britain's Society for Popular Astronomy (SPA) since 1984, is the Assistant Director of the Lunar Section of the British Astronomical Association (BAA). He edits and produces three astronomy publications – Luna (journal of the SPA Lunar Section), The BAA Lunar Section Circulars and Popular Astronomy Magazine. He is also layout editor for the Bulletin of the Society for the History of Astronomy.

Grego's astronomical writings and observations have featured in many publications since 1983, including the BAS Newsletter, Popular Astronomy, The New Moon, Amateur Astronomy and Earth Sciences, Gnomon, The Lunar Observer, Yokohama Science Center News and the CD-ROM Window on the Universe. Since 1997 he has written and illustrated the monthly MoonWatch page in UK's Astronomy Now magazine, and he is the observing advisor and columnist for Sky at Night magazine.

He has given many talks to astronomical societies around the UK and has featured on a number of radio and television broadcasts.

Grego is the author of numerous astronomy books, including "Collision:Earth!" (Cassell, 1998), "Moon Observer's Guide" (Philips/Firefly, 2004), "The Moon and How to Observe It" (Springer, 2005), "Need to Know? Stargazing" (Collins, 2005), "Solar System Observer's Guide" (Philips/Firefly, 2005), "Need to Know? Universe" (Collins, 2006), "Exploring the Earth/Exploring the Moon/Discovering the Solar System/Voyage Through Space/Discovering the Universe" (five book in the QED Space Guides series, 2007), "Venus and Mercury and How to Observe Them" (Springer, 2008), "The Great Big Book of Space" (QED, 2010), "Galileo and 400 Years of Telescopic Astronomy" (Springer, 2010) and "The Star Book" (D&C, 2012).

Grego maintains his own website at www.lunaobservers.com (which occasionally features live webcasts of the Moon and planets and other astronomical phenomena) and is webmaster for the BAA Lunar Section at www.baalunarsection.org.uk.

He is a member of the SPA, ALPO, SHA, and BAA and is a Fellow of the Royal Astronomical Society.

Contents

Contents

Part I

Our Current Knowledge of Mars

Chapter 1

Fourth Rock from the Sun

1.1 Physical Dimensions

Compared with most of the Solar System's planets, Mars is a little on the small side. Measuring an average of 6,792 km in diameter, its dimensions lie neatly between that of the Earth and the Moon, making it the Solar System's seventh smallest planet. Its surface area of 144,798,500 km² – some 28% of the Earth's surface area – is roughly equivalent to the area of dry land on the Earth.

Mars in comparison with the Earth (Credit: NASA/Grego)

P. Grego, *Mars and How to Observe It*, Astronomers' Observing Guides,
DOI 10.1007/978-1-4614-2302-7_1, © Springer Science+Business Media New York 2012

In certain respects Mars has similarities with the Moon, with one heavily cratered hemisphere contrasting with another hemisphere which is relatively smoother. Yet Mars is also the most Earth-like planet in the Solar System, with its seasons and atmosphere, familiar weather phenomena such as dust devils, dust storms and various types of clouds. There is much physical resemblance to the Earth, too, with abundant evidence of past rivers and oceans, and wind-formed features similar to those of our own planet. And, perhaps, lowly forms of life evolved on Mars – and the planet may still be an abode of life.

Like the Earth, Mars is not a perfect sphere, but it is slightly oblate, with an equatorial bulge and flattening at the poles – a shape, actually shared by all the planets, which is produced by the planet's axial rotation. Measured across the equator, Mars is 6,792 km across, while from pole to pole it's 40 km smaller, at 6,752 km. In terms of volume, Mars is 15% the size of the Earth – some 160 billion cubic kilometers compared with Earth's 1.1 trillion cubic kilometers – and therefore six Mars globes could fit inside one the size of the Earth and still leave room to spare.

Mars is not a simple flattened globe – it's slightly ellipsoidal. We've already noted that its northern and southern hemispheres appear markedly different, and this is brought home by the fact that there's a 3 km difference between the planet's center of mass and its center of figure. The boundary between the two hemispheres is characterized by broad, gradually sloping plains spreading over thousands of kilometers across the boundary. In addition, localized areas such as the giant volcanic Tharsis bulge straddling the planet's equator – an uplifted continental mass equivalent to the size of the continental USA – rises several kilometers above the mean Martian surface level. On the opposite side of the planet lies a somewhat less-pronounced bulge, Arabia Terra, a heavily cratered region thought to be one of the most ancient terrains on Mars.

1.2 Mass, Density and Gravity

Although Mars' volume is 15% that of the Earth, its mass is just 10.7% the Earth's, which means that Mars as a whole is made of lighter material than our own planet. On average, Mars stuff is 3.3 times denser than water, while Earth stuff is 5.5 times denser than water. With its small size and low density, Mars has a surface gravitational force just 38% of that found at the surface of the Earth. On Mars, an Olympic power lifter could easily lift an Earth weight of 1 t, as it would weigh only 377 kg on Mars, and people would feel 62% lighter than on Earth; an object dropped by an astronaut would fall to the planet's surface noticeably more slowly than the same rock would fall to Earth.

1.3　Orbit

Mars, the fourth and furthest terrestrial planet from the Sun, has an orbit lying entirely between the orbits of the Earth and Jupiter. At an average distance of 227.9 million kilometers from the Sun and traveling at an average speed of 24 km/s, Mars completes one orbit around the Sun (with respect to the stars) every 1.88 years (687 Earth days) – the length of the Martian year. The planet's synodic period – the time it takes to appear in the same position relative to the Earth and Sun – amounts to 2.1 years (780 Earth days).

Mars' orbit is inclined by just 1.9° to the plane of the ecliptic, but it is one of the Solar System's least circular planetary paths; with an eccentricity of 0.09, its orbit is second only to the more eccentric orbit of Mercury. Mars' orbit takes it from a perihelion of 206.7 million kilometers (1.38 AU) to an aphelion of 249.2 million kilometers (1.67 AU) – a difference of 42.5 million kilometers between its perihelion and aphelion distances – an average orbital distance 78.3 million kilometres further than that of the Earth.

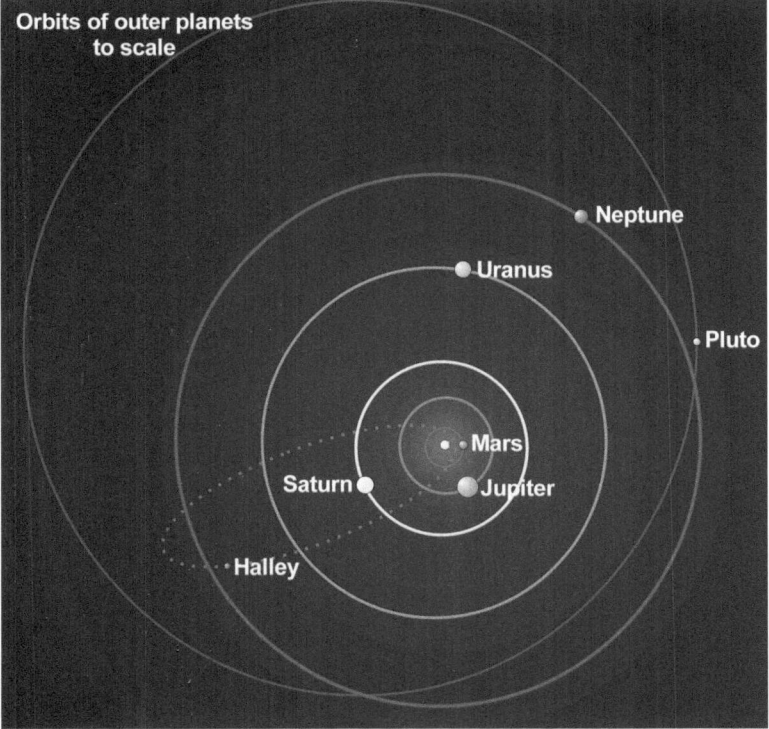

Orbits of the outer planets (Credit: Grego)

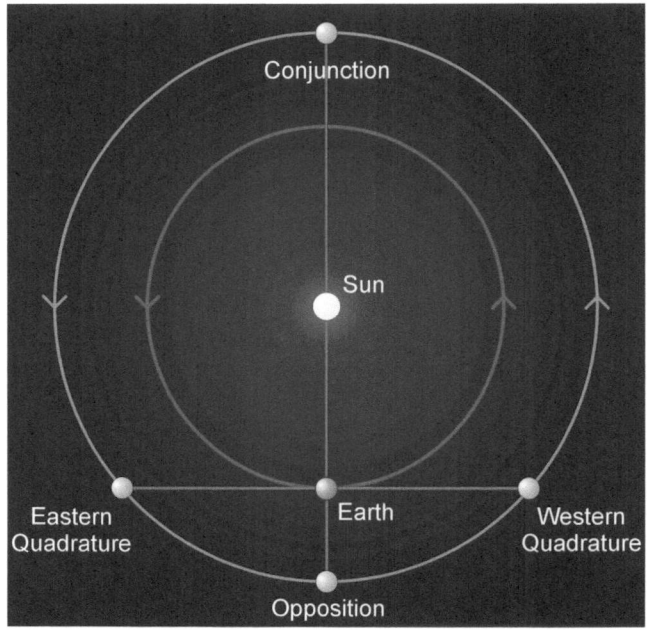

Orbital phenomena displayed by Mars and the outer planets (Credit: Grego)

1.4   Axial Rotation and Inclination

Rotating on its axis once every 24.6 h, a Mars day, or sol, is just 39 min 35 s longer than an Earth day. 669 sols make up one Martian year, and each orbit of Mars around the Sun sees it rotate on its axis 669 times. Mars has an axial tilt of 25.2°, similar to the Earth's 23.5°. So, while we have Polaris (Alpha Ursae Minoris, magnitude +1.97) as our familiar North Star less than 1° from our north celestial pole, Martian skies host the somewhat brighter Deneb (Alpha Cygni, magnitude +1.25) near its north celestial pole. Because of its axial tilt, Mars experiences a cycle of seasons, just like the Earth, only they last around twice as long. Southern spring (northern autumn) on Mars lasts 146 days; southern summer (northern winter) goes on for 160 days; southern autumn (northern spring) is 199 days long; and southern winter (northern summer) lasts for 182 days.

Mars' obliquity (its axial tilt) changes from around 15° to 35° in cycles of 120,000 years, these changes accruing through the gravitational influence of other planets; the planet is currently in the middle of an obliquity cycle. It has been speculated that at intervals of several tens of millions of years Mars' obliquity may range between 0° and 60° so that sometimes the planet effectively orbits the Sun 'on its side' (as, currently, does Uranus). Earth's obliquity only varies between 22.1° and 24.5° over a 41,000 year period, due to the stabilizing influence of the Moon.

Increased obliquity sees higher levels of sunlight reaching the Martian polar regions, causing more water at its poles to sublimate and enter the atmosphere; this water vapor condenses as ice or snowfall in the cooler equatorial regions. At its highest obliquity more carbon dioxide is given off by the thawing ice caps, and as a result the atmospheric temperature and pressure increases to a point where liquid water can exist on the planet's surface. There's ample evidence for such cycles of thawing and freezing on Mars.

Chapter 2

History of the Red Planet

2.1   Formation

Four large protoplanets composed largely of silicates and metals – Mercury, Venus, Earth and Mars – had grown to dominate the inner Solar System within around 200 million years of the collapse of the solar nebula and the formation of the protosun. None of these protoplanets is thought to have been massive enough to have attracted a disk of material from which its own satellite could have formed.

The protoplanets of the inner Solar System swept up all the available material within their own orbits, attracting dust and gas and accumulating large amounts of matter through countless impacts. High temperatures within the planets arose as a result of asteroidal impacts, internal pressure and the radioactive decay of elements, and as a consequence there was period of melting of the material making up the protoplanets. Consequently, a process of differentiation took place within the molten protoplanets as heavier elements sank to form their iron-rich cores, while lighter material rose to form their mantles and crusts.

Current models of the dynamics of the early Solar System show that the initial angular momentum of the solar nebula is reflected in the angular momenta of the planets and their orbits. A planet's rate of rotation and axial inclination can change because of tidal gravitational interactions between it and the Sun, and also through major asteroidal impact. Several very large impacts took place early on in Mars' history.

As the crusts on the early Solar System's protoplanets thickened and consolidated they began to retain the imprints of countless asteroidal impacts in the form of craters and basins. Lava flows intruded through crustal fissures, filling the floors of many of the impact features. Clearly visible remnants of an intense period of asteroidal impacts called the late heavy bombardment (which ended about 3.8 billion years ago) can be seen on Mars, in addition to the Moon and Mercury – those on Venus and the Earth have long since disappeared due to widespread vulcanism and tectonics. Some of these impactors came from the inner Solar System, but many originated in the outer Solar System, diverted towards the Sun after gravitational interactions with the giant outer planets.

P. Grego, *Mars and How to Observe It*, Astronomers' Observing Guides,
DOI 10.1007/978-1-4614-2302-7_2, © Springer Science+Business Media New York 2012

2.2 Surface History

After its formation and the late heavy bombardment Mars cooled relatively quickly, which meant that plate tectonics had no opportunity to get started. Today's Martian surface comprises four general types of terrain – ancient, relatively intact terrain rich in craters, younger volcanic plains, large volcanic structures and swathes of sedimentary deposits.

Terrestrial geologists use stratigraphy – the study of rock layers (strata), placing them into a relative historical context – to decipher much about the history of the Earth's crust. Stratigraphy makes use of the law of superposition, which states that older layers of rock are overlain by younger layers. However, a cross-section through virtually any part of the Earth's crust will show that the stratigraphic picture isn't nearly as straightforward as a simple layer-cake. A number of processes – chief among them crustal movement, folding, faulting and erosion – create a complex picture. For example, older rock layers may be folded over to overlay younger ones, and whole sequences of rock deposition may be missing altogether from the record. In addition to determining the relative ages of strata, terrestrial geologists can determine absolute rock ages by radiometric dating, where naturally occurring radioactive isotopes in rock (which decay at a known rate) are compared with the abundance of their decay products.

Planetary geologists don't have nearly as much evidence to base their findings on when applying stratigraphy to unravel the history of the surface of Mars – there are no geologists on the ground and no fully-equipped laboratories in which to analyze the Martian rocks. But science has made a start. Increasingly detailed images and data from space probes in Martian orbit and investigations carried out from several points on Mars' surface have enabled planetary geologists to make many firm conclusions about the history of the planet, in addition to providing many tantalizing clues about the sequence of events that took place on the surface in local and global contexts. Superposition of topographical features, such as craters, faults, valleys, igneous and sedimentary deposits, can be used to determine their relative ages.

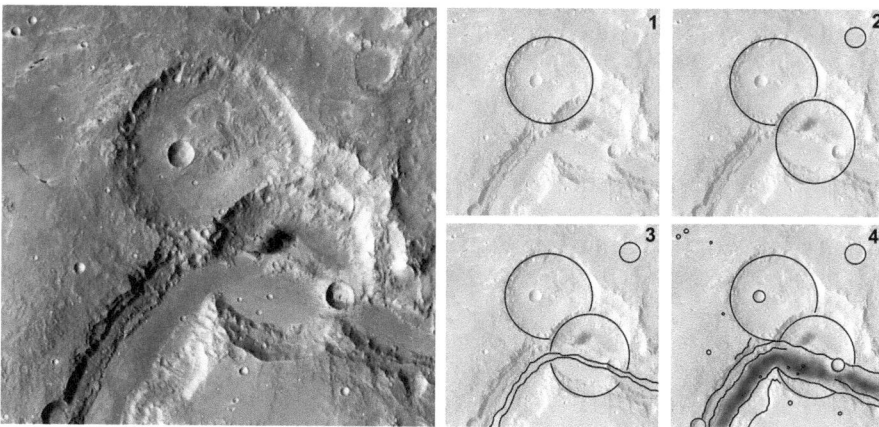

A great deal can be gleaned by applying superposition to topographic features, determining the sequence in which Martian features formed. Taking an example, as shown in this image, the plain to the southeast of Gusev crater shows a clear stratigraphical order of overlapping features. The oldest obvious topographical feature is a 30 km impact crater (1); this, in turn, has been overlain in part by another, similar-sized impact crater (2), a feature which has then been cut across by a winding valley, formed as a result of running water (3). As the river valley broadens and sediment is deposited on its floor, smaller impacts dot the landscape, some of them overlapping the walls of the valley (4) (Credit: NASA/Peter Grego)

Where impact craters are well-preserved on a planet's surface – as in the case of the Moon and Mars – planetary geologists can also use counts of crater density to determine the relative ages of individual features, such as crater floors, as well as broader geographical areas. As we've seen, Mars' northern hemisphere is far more densely cratered than its southern hemisphere, and on the whole its craters are larger than those in the south; this strongly suggests that the surface in the north is much older than that in the south. However, the southern hemisphere is likely to have experienced just as much asteroidal bombardment as the northern hemisphere during the late heavy bombardment, but the southern hemisphere has been extensively resurfaced and reshaped by a variety of processes, including volcanism and the action of water. In order to build up a reliable relative timescale of the planet's surface, based on crater counts, it's necessary to have a good idea of the rate of impact crater flux (the rate of formation by crater size per unit area) over geological time. Counts on Mars have been based on the better-known impact crater flux observed on the Moon, a body whose surface displays a well-preserved record of impact.

2.3 Crater Density Timescale

A timescale based on Martian stratigraphy suggests that Mars experienced an early warm period, with major asteroidal bombardment; this was followed by rapid cooling, extensive vulcanism and periods of climate change which determined the way in which water was expressed in the crust and on the surface. The planet's history can be separated into four distinct periods, each named after large surface features thought to have formed during these periods.

1. **Pre-Noachian Period** (4.5 to 4.1 billion years ago). Early in this period saw the development of the Martian dichotomy. Topographical and gravitational studies from orbit have revealed a striking difference in crustal thickness between the two hemispheres, averaging 32 km thick in the north and 58 km in the southern highlands. It is widely thought that the Martian dichotomy may have been produced by a single huge impact in the north, the impactor measuring up to 60% the diameter of our own Moon, creating the Borealis Basin. If this feature were of impact origin it would be the largest impact basin in the Solar System, measuring 10,600 km long and 8,500 km wide. Such a catastrophic impact will have blasted away a great portion of the early Martian atmosphere. It has been suggested that several large early impacts may have produced the Borealis Basin, but those large (and mainly buried) impact structures which have been identified in the north show signs of having post-dated its formation. The possibility that the dichotomy was produced by purely endogenous processes, where a particular form of mantle convection (known as degree-1 convection) produced upwelling in the southern hemisphere and downwelling in the north, is not supported by conventional mantle convection models.

   The Pre-Noachian Period saw the asteroidal battering of the late heavy bombardment, a cataclysm that affected all the terrestrial planets of the inner Solar System, including the Moon. Crater density studies of the 20 or so 1,000 km-plus large Martian impact basins suggest that most of them were produced in an astronomically short period of time; 18 of them may have formed within the first 200 million years, including the Hellas basin (2,300 km across) in the southern hemisphere and the Argyre and Isidis impact basins (1,800 and 1,500 km across, in the southern and northern hemispheres respectively). Further impacts, erosion and resurfacing have obliterated many of the features formed during this period. The last few tens of million years of the Pre-Noachian Period also saw the demise of Mars' global magnetic field, an event which has been linked with heavy asteroidal bombardment. Following the magnetic field's shutdown, the solar wind would have stripped the atmosphere; unimpeded in reaching the surface the solar wind further impaired the conditions suitable for the development of life.

2. **Noachian Period** (4.1 to 3.7 billion years ago). The oldest surfaces visible on Mars were formed during the 'wet and warm' Noachian Period. During this period the Tharsis bulge, a roughly circular, continent-sized elevation formed over a mantle hotspot straddling the Martian equator (spanning around 120° in latitude and 80° in longitude, centered at around 95°W). The giant Hellas impact basin was also formed during this period. There is abundant evidence that liquid water covered large areas of Mars' surface during the Noachian period, forming lakes with erosional outflow valleys. The warmer, wetter Mars of the middle-Noachian may have hosted an extensive oceanic tract.

Towards the late Noachian Period, as water was lost to the planet by impacts and atmospheric sputtering, there came a major change in the Martian climate as the water table lowered deep beneath the surface. There were localised areas where groundwater upwelled and evaporated, for example the playa evaporates (dry salt lake beds) found in Meridiani Planum and Arabia Terra. Periodic short-term changes in Mars' orbit altered the amount of water present in the polar caps and cryosphere, which explains the periodic sedimentary layering and erosional unconformities (producing gaps in the sedimentary record) found on the planet, for example in Meridiani Planum.

3. **Hesperian Period** (3.7 to 3 billion years ago). Extensive lava plains erupted onto the Martian surface during the Hesperian Period, along with the start of the formation of Alba Mons and Olympus Mons, the largest shield volcano in the Solar System. Rising atmospheric pressure and temperatures caused catastrophic floods of water which carved the planet's large outflow channels around Chryse Planitia, emptying into temporary lakes in low-lying topography in the north.

4. **Amazonian Period.** (3 billion years ago to the present day). Dwindling volcanic activity, along with glacial activity and occasional releases of liquid water has occurred during the Amazonian Period. Landscapes of this period are the youngest on Mars and are characterised by a relative paucity of impact craters.

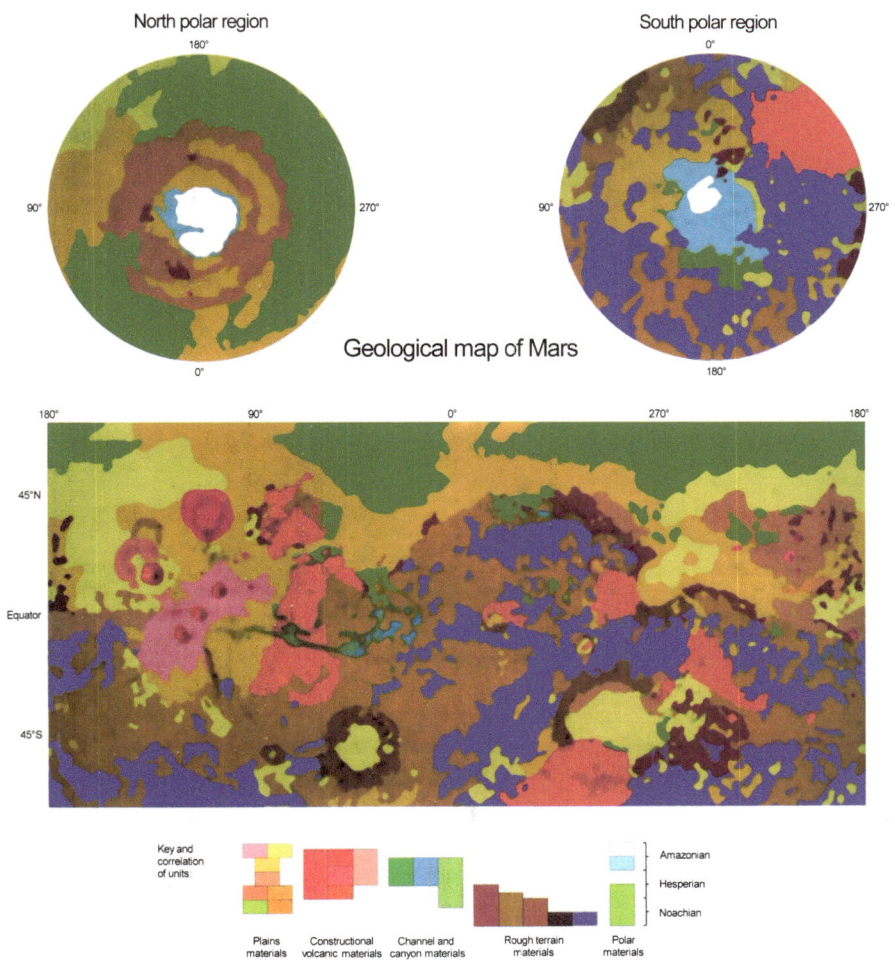

North polar region

South polar region

Geological map of Mars

Key and correlation of units:

Plains materials
Constructional volcanic materials
Channel and canyon materials
Rough terrain materials
Polar materials

Amazonian
Hesperian
Noachian

Geological map of Mars (Credit: NASA/USGS/Grego)

History of the Red Planet

13

Chapter 3

Stuff and Substance

Like the Earth, Mars is a terrestrial planet made chiefly of silicates (minerals containing silicon and oxygen), along with metals and a variety of other elements that compose igneous rock. To determine the interior composition of Mars scientists require information on several parameters; these include the planet's total mass, its size, the moment of inertia and core mass; the latter has not yet been scientifically established. We do know that Mars, like the Earth, has differentiated contents, with a dense central metallic nickel-iron core surrounded by a silicate-rich mantle and overlain by a solid rocky crust. Mars' mantle is twice as rich in iron as that of Earth, and it also contains a greater proportion of potassium and phosphorus.

Mars' core mass is unknown, but it likely accounts for between 6% and 21% of its overall mass; in terms of its dimensions, lack of certainty about its mass and composition means that it is probably somewhere between 2,600 and 4,000 km in diameter, but it's not known whether the core has solid or liquid components. In comparison, the Earth's nickel-iron core is around 7,000 km across with both a central solid and outer liquid component; the enormous pressure causes the core's central solidity, while high temperatures in the outer core may melt the iron. If Mars' core is mainly composed of iron, then its maximum diameter would be around 2,600 km; if the core contains a large amount of sulphur, its diameter would be required to be considerably larger, up to 4,000 km across.

Most of the underlying Martian crust and a large proportion of its surface is composed of basaltic material, similar to tholeiitic basalt which makes up Earth's oceanic crust and the lava plains of the Moon. In addition to silicon and oxygen, Mars' crust contains iron, magnesium, aluminium, calcium and potassium plus smaller proportions of titanium, chromium, manganese, sulphur, phosphorus, sodium, and chlorine. Volatile elements such as sulphur and chlorine are present in higher proportions in Mars' crust than that of Earth.

Hydrogen exists in the crust and regolith of Mars in the form of water ice and in hydrates – minerals that have been chemically altered by water to form a new mineral (either through the transformation of oxides into hydroxides or when water seeps into a mineral's crystalline structure, producing clay minerals).

Carbon is present on the surface in the form of dry ice at the poles, in small proportions in Martian dust, and in localized outcrops of carbonates. Carbonates form only in water of a relatively high pH value, from very mildly acidic (a pH of 5) to alkaline (greater than a pH of 7). Terrestrial seawater has a pH value of

P. Grego, *Mars and How to Observe It*, Astronomers' Observing Guides,
DOI 10.1007/978-1-4614-2302-7_3, © Springer Science+Business Media New York 2012

around 8, and carbonates are introduced into the oceans by rivers and along vents in the oceanic ridges. Carbonate oozes (biogenic sediment) blanket around half of Earth's sea floors, but they are only found above a depth of 4,500 m because carbonates dissolve at greater pressures. Examples of common terrestrial carbonates include limestone and chalk. Carbonates provide evidence of the presence of liquid water on Mars' surface at some point in the past, but it was only relatively recently that planetary geologists succeeded in discovering Martian carbonates through spectral analyses conducted from orbit and on the surface.

Small amounts of carbonates, particularly magnesium carbonate, had been detected in the Martian surface dust in 2003, but the mineral's widespread distribution in the dust across the surface was not suggestive of any specific sources. Orbital studies in 2008 revealed exposed carbonate outcrops in a number of areas. The first of these to be discovered lay in 10 km^2 of the Nili Fossae region in Mars' northern hemisphere, an arcuate fracture valley on the northwestern border of Isidis Planitia. Here, around 3.6 billion years ago, igneous rock (possibly in the form of dykes) appears to have been hydrothermally altered to form magnesium carbonate rocks. The discovery also threw up new hopes of finding fossilized life in this region, as it proves that waters flowing through Nili Fossae were not acidic and were therefore conducive to life. Indeed, the site is similar to the Pilbara outcrops of northwestern Australia, where evidence for ancient fossilized life up to 3.5 billion years old remains evident to this day in the form of small, dome-like features called stromatolites, made up of successive layers of microbe communities.

Carbonate rocks show up as green in this image of a portion of Nili Fossae (Credit: NASA/MRO)

In 2010 the Spirit rover, while trundling across in the Columbia Hills inside Gusev crater, found rock outcrops at Comanche and Comanche Spur which were rich in iron-magnesium carbonate. The Columbia Hills represent an 'island' of older (Noachian era) rocks surrounded by younger (Hesperian era) volcanic rocks; the carbonates at Comanche are likely to have precipitated under hydrothermal conditions from near-neutral pH carbonate-bearing water.

Comanche (mid-ground, *left*) and Comanche Spur (mid-ground, *right*), where Spirit discovered iron-magnesium carbonates (Credit: NASA/JPL)

Calcium carbonate has been found on the surface in the planet's arctic soil; comprising up to 5% of the soil's weight it's likely to have formed there through the interaction of carbon dioxide in the atmosphere with films of liquid water on the surfaces of particles in an alkaline soil.

At Meridiani Planum the Opportunity rover discovered an area whose sulphate salt evaporate rocks contained spherules of the iron mineral haematite and a surface and regolith littered with loose spherules which had been eroded out of the rocks by sandblasting. Known as 'blueberries' because of their bluish hue in false color images, their origin remains uncertain. It was initially speculated that the spherules were formed in sprays of molten rock thrown out by vulcanism or impact, in which case they would likely be distributed in layers. However, it was discovered that the spherules are distributed evenly and randomly in the rocks, suggesting that they formed in-situ as a result of concretion in a liquid water environment.

Spherules of haematite imaged the Mars Exploration Rover Opportunity (Credit: NASA/MER)

3.1 Winds of Change

Mars' distinctly red hue, a tint easily noticeable with the unaided eye, is due to iron oxide (rust) in the upper layers of the planet's regolith. Although the loose surface material is frequently and generically referred to as 'sand', most of it is actually much finer material, closer to the geological definition of dust (particles between 10 and 50 μm across).

Early telescopic observers soon discovered that the Red Planet's north and south hemispheres differ markedly in appearance. Through the telescope, extensive dusky tracts cover Mars' southern hemisphere, while the northern hemisphere appears much brighter, with a few isolated dusky patches. Although considered permanent, those telescopically familiar light and dark features appear to vary in outline and intensity from one season to the next, but they do not always correspond with Martian topography. Dark areas such as Syrtis Major and Mare Acidalium (to use the old-style nomenclature favoured by amateur observers) are darker tracts of the surface whose apparent shape and intensity varies on a seasonal basis because of transient coverage by wind-blown dust of a lighter tone. This is particularly evident in space probe images showing streaky accumulations of fine, light colored dust forming in thin layers leeward of elevated crater rims, indicating the direction of the prevailing winds. Forming as a result of soil creep under the influence of low winds during large Martian dust storms, such streaks downwind of crater rims and other elevated surface formations are the most common of Mars' wind-formed features.

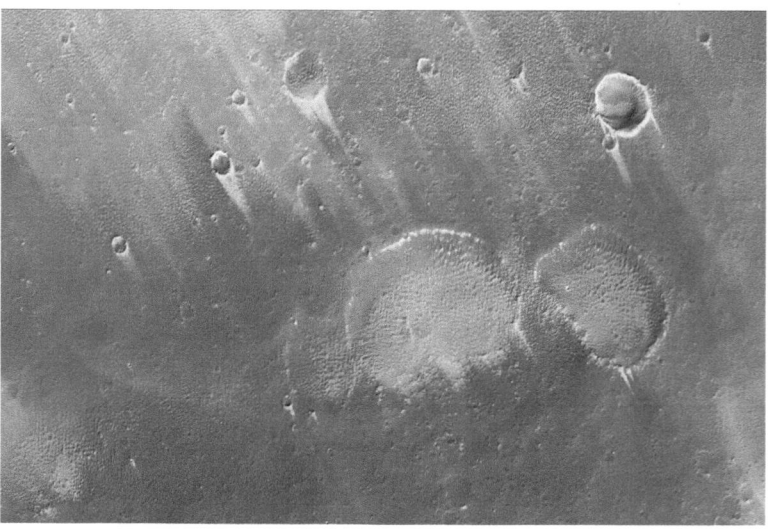

Bright wind streaks formed on the southwest slopes of the Martian volcano Pavonis Mons (Credit: NASA/JPL/Malin Space Science Systems)

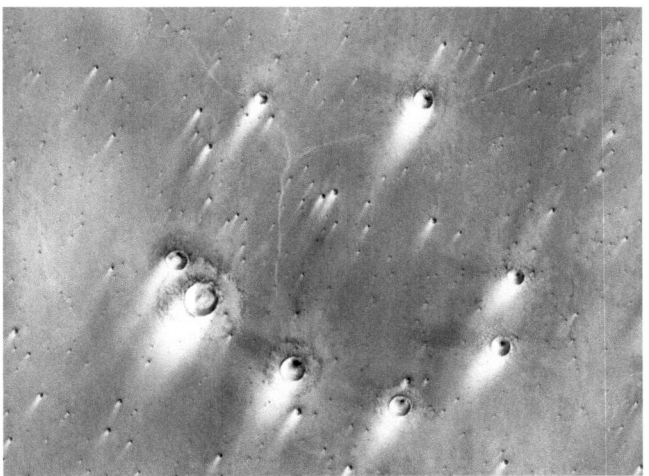

A group of small craters with bright wind streaks in Chryse Planitia (Credit: NASA/JPL/Malin Space Science Systems)

Wind streaks made up of larger sand-sized particles can form following large dust storms by saltation, a process where high winds lift soil particles off the surface and redeposit them at a distance downwind. This type of dark streak formation is vividly seen in areas where thinning water ice coverage during summertime at the poles exposes patches of the darker underlying surface; high winds disturb the exposed soil and dark material is distributed downwind, forming dark fans. Another mode of dark streak formation is avalanching, where lighter surface material slides down steep surface slopes, exposing a darker underlying surface and giving rise to dark paper tear-like streaks; over time the streaks have been observed to lighten in tone and fade. Temporary dark streaks also form from the scouring effects of Martian dust-devils; images show their twisting paths across the surface.

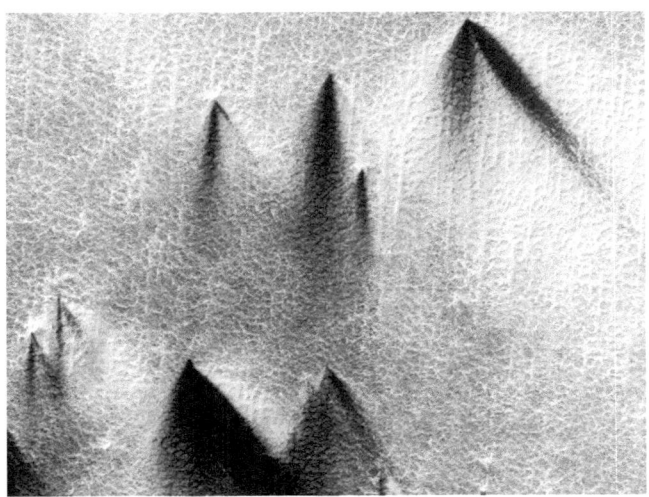

These dark fan-shaped streaks emanate from exposed patches of soil in the retreating south polar ice cap (Credit: NASA/JPL/University of Arizona)

These dark streaks run downslope in Acheron Fossae (Credit: HiRISE/MRO/LPL/NASA)

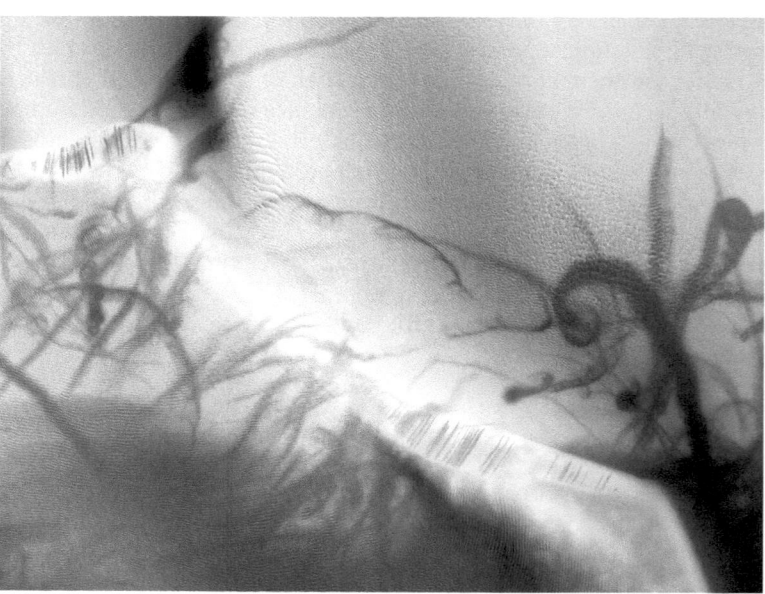

Dark swirling patterns on the Martian surface mark the paths of dust devils. Note also the linear dark streaks on the slope of the scarp, formed by avalanching (Credit: HiRISE/MRO/LPL/NASA)

Dunes, often similar in size and shape to those which form in the Earth's deserts, are commonly found on Mars, both on the planet's exposed plains and on many crater floors. Extensive dune fields can be found in the north circumpolar region, where dunes of many different types have been identified. Formed by the action of wind on loose surface material, dunes can change in appearance over time. Depending on local topography, dunes can slowly migrate across the surface or are held in check by local topography – trapped within craters or held in place by other obstacles – and nothing short of a change in wind strength and/or direction is capable of shifting them.

Transverse dunes – also known as transverse aeolian ridges (TARs) – take on a number of forms, from simple ripples to long and linear ridges, forming at right angles to the direction of the prevailing winds which fork at their ends or join up with neighboring TARs in low-angle Y-shaped junctions. TARs, more common in the southern hemisphere than in the north, are generally found in a belt between 30° N and 30°S and are abundant in the Meridiani Plaunum region and within southern craters. TARs found near layered terrain are millions of years old, while those adjacent to large dark dunes (LDDs) are more recent formations.

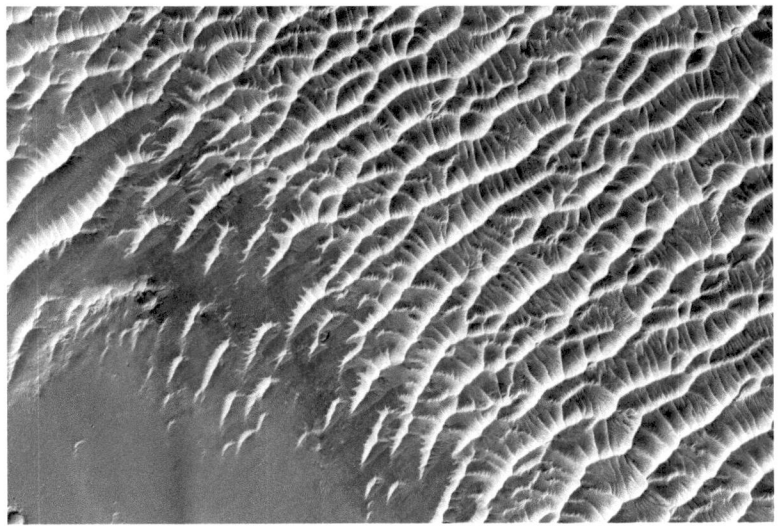

Transverse aeolian ridges near the crater Schiaparelli in Mars' equatorial region of Mars. With a southwest-to-northeast trend, these features formed from winds blowing from the northwest or southeast. The image covers an area about 750 m across (Credit: NASA/MGS)

LDDs are odd looking features with low albedo (less than 0.15), rounded and slug-like in appearance with wavelengths of a few hundred to a few thousand of meters. Composed of dark basaltic sand with a finer particle size than TARs, they are recent features that overlie terrain of lighter colored material.

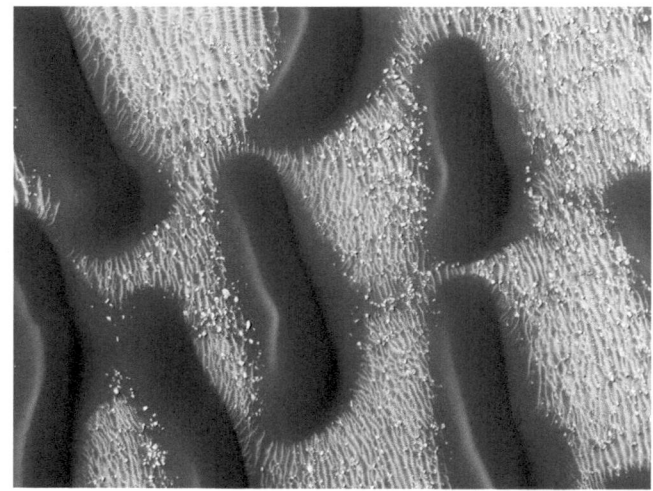

A field of large dark dunes (LDDs) in the crater Proctor, formed by westerly winds of the Martian autumn and winter. Note the much smaller TARs on the ground between the LDDs (Credit: NASA/JPL)

Barchan dunes, similar in size and shape to terrestrial dunes of the same name, are streamlined crescent-shaped dunes with curving horns that point downwind. Sand blown over barchan dunes falls on their leeward slip face, between the horns, so that over time the dune migrates downwind across the landscape in the direction of its horns.

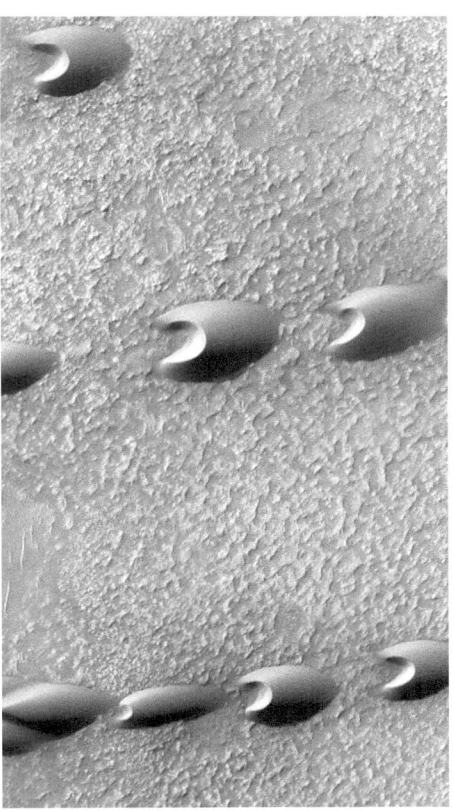

A cluster of horseshoe-shaped barchan dunes within the Hellespontus region in Mars' southern hemisphere. Their shape and orientation indicate a west to east (right to left) wind direction (Credit: NASA/MRO)

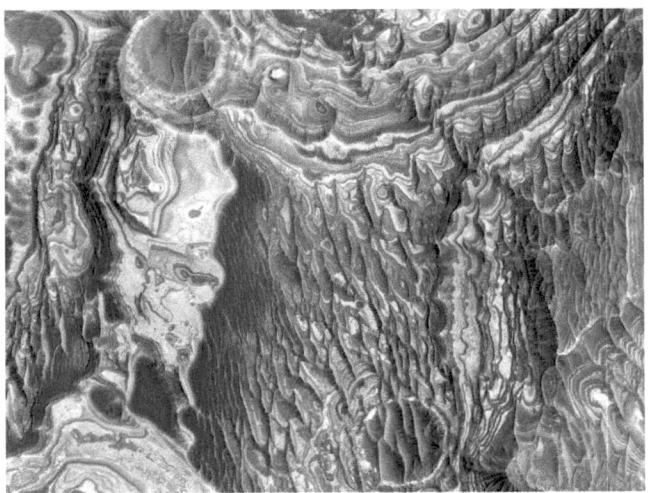

Shaped by wind, water, lava flows, seasonal icing and other forces over the aeons, the surface of Mars shows some dramatic sights. This remarkable view shows colour variations in bright layered deposits on a plateau near Juventae Chasma in the Valles Marineris region. That these bright layered deposits, partly covered by a brown mantle, contain opaline silica and iron sulphates (Credit: NASA/JPL/LPI)

Wind erosion can form a variety of landscape features. In landscapes made up of soft, easily-eroded material (such as consolidated sedimentary deposits or volcanic ash), grooves and ridges parallel to the prevailing winds can be produced by dust and sand abrasion; most grooves are symmetrical and have V-shaped profiles. Where strong unidirectional winds scour a surface made of material of variable hardness, erosion occurs more rapidly on softer material. As the landscape changes and the winds are channeled to produce characteristically eroded landscapes, yardangs are formed from harder material – hills and mounds with a streamlined appearance. In their advanced state yardangs are commonly about three to four times longer than their width and bear a resemblance to the upturned hull of a boat, with a steep, broad blunt end facing the wind and a shallower pointed end facing downwind (with respect to the winds that originally formed them). Yardangs often display unusual forms, especially where deflation has caused turbulent eddies to erode in specific places around the feature's base.

Wind grooves score the lower southern slopes of Olympus Mons (Credit: ESA/Mars Express)

Yardangs in southern Amazonis Planitia (Credit: NASA/JPL)

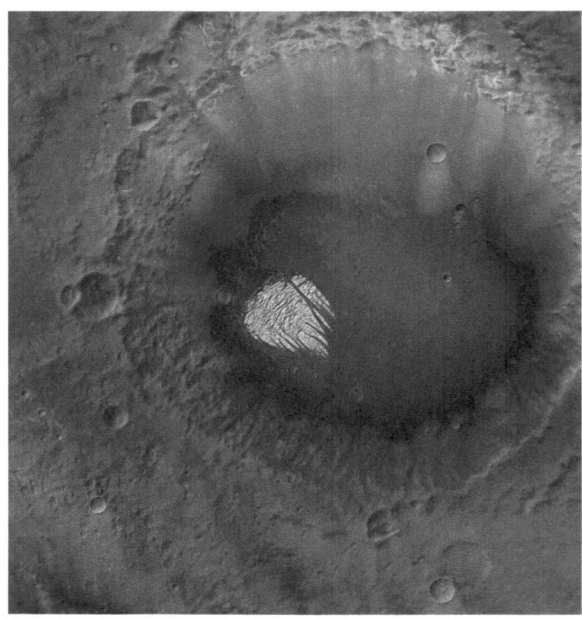

Inside the 95 km crater Pollack lies an unusual ridged mound known as 'White Rock', a layered sedimentary outcrop 15 km wide. Aeolian erosion has created steep cliffs and valleys across 'White Rock', while dark overlying dunefields on the crater floor accentuates its lighter tone (Credit: ESA/Mars Express)

3.2 Fluvial Erosion

As the young Mars cooled through a reduction of heat flow from the planet's interior and changing surface conditions, the planet's water-rich regolith formed a global cryosphere several kilometers thick. As a result, water contained in underlying porous material was pushed down against deeper, less porous material. Owing to the peculiar variations in the altitude of the Martian surface – generally high altitudes in the southern hemisphere and low altitudes around much of the equator – the frozen ground is around 6 km deep around the south polar region and just 2.5 km deep in the equatorial regions.

Various processes have breached the cryosphere during the planet's history, including igneous intrusions, tectonic processes and large impacts, leading to pressure-driven outbursts of liquid water at the surface. The most dramatic features are outflow channels and valley systems – features formed by the erosive action of water following the melting of underground reservoirs of ice. As the water flowed down-slope across the landscape, simultaneously freezing and evaporating along its length, blocks of ice and debris increased the erosive action of the floodwaters. Although parts of the flow might have frozen over at the surface, it continued to flow beneath, similar to terrestrial eskers or frozen rivers. Some outflow channels display tributary channel systems and streamlined bedforms such as 'teardrop islands' or mesas, formed when raised rocky crater walls channeled floodwaters around them leeward of the flow direction.

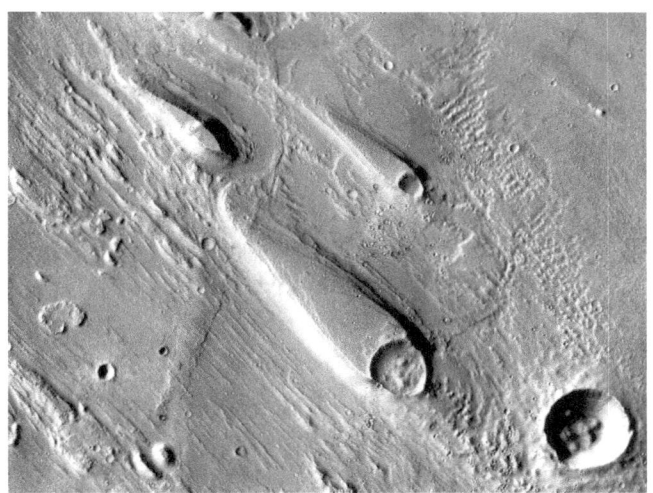

Teardrop-shaped 'islands' formed by floodwaters can be seen in the lee of these craters in the Ares Vallis region, along with terrain striation, all indicative of floodwaters flowing from lower right to upper left (Credit: NASA/JPL)

There's a great range in size of Martian outflow channels; some originate from sizeable regions, many hundreds of square kilometers in area, while others have smaller, more localized origins. Most of Mars' outflow channels features lie in the planet's equatorial regions in the northern lowlands.

A preponderance of large outflow channel origins can be found in regions adjacent to Chryse Planitia, a relatively smooth plain some 1,600 km across in the planet's northern equatorial region adjacent to Lunae Planum in the west; an ancient impact basin, Chryse Planitia has a floor 2.5 km below the planet's mean surface level. Some 3,000 km long and 230 km across at its widest, Kasei Valles is Mars' largest outflow channel. Like other large outflow channels, it suddenly begins, at its widest, in a region of chaotic terrain and develops striking river valley landforms (notably of low sinuosity) along its length, including anastomosing forms (channels that split and then reconnect further along their path) before entering its terminal basin. At their terminal ends, sedimentary deposits and landscape features typical of terrestrial outflow systems (such as deltas) are lacking in those of Mars; instead, they appear fade into the landscape, a consequence of pooling of water in their terminal basins.

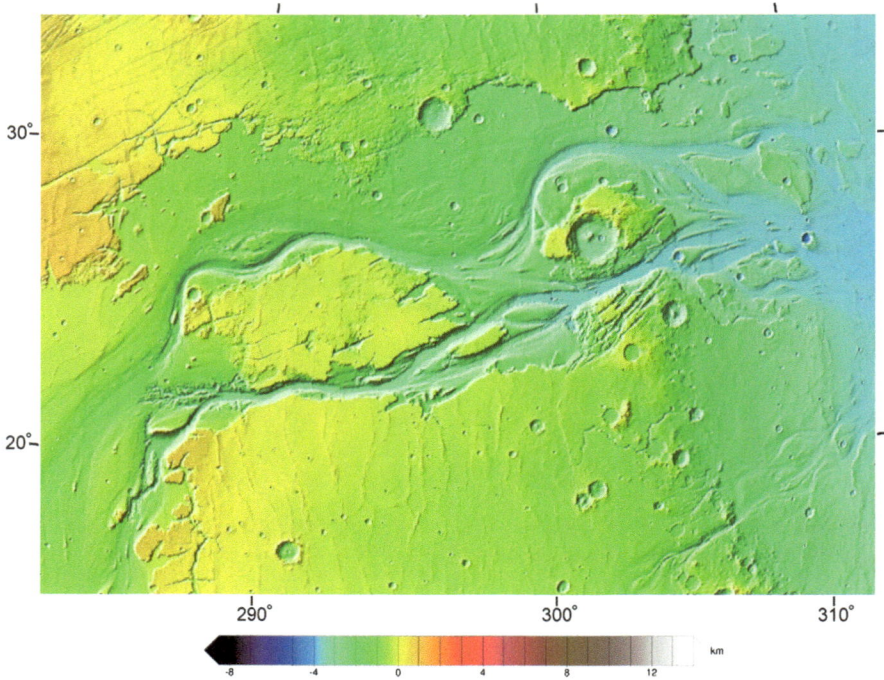

Topographic map of Kasei Valles, Mars' largest outflow channel. It begins at left in a region of chaotic terrain on the eastern slopes of Tharsis and heads west towards the lowlands of Chryse Planitia at right (Credit: NASA)

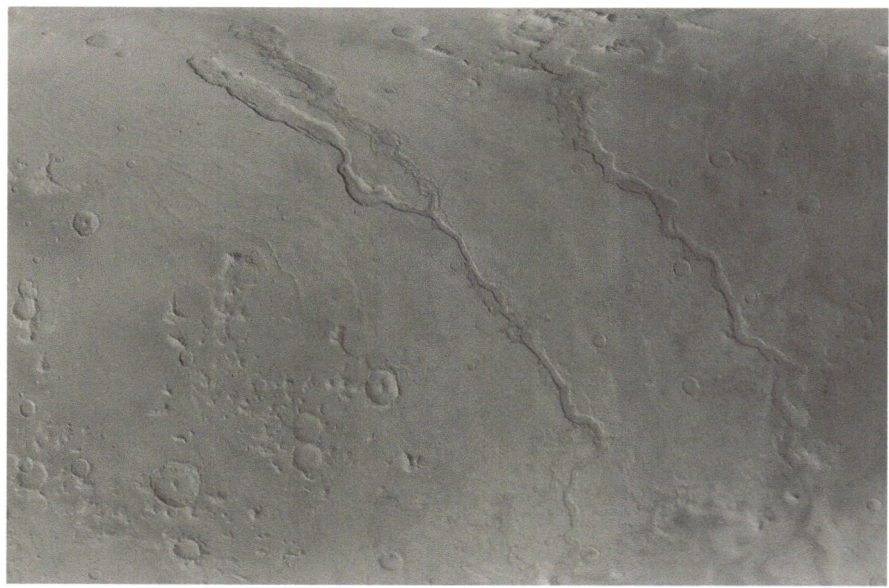

Outflow channels Dao Vallis (upper left) joins Niger Vallis (just above centre), with Harmakhis Vallis (at right). Each was formed by water flowing from top to bottom in this image, which covers an area 800 km across, north is toward the left (Credit: NASA/MOC)

Most of the outflow channels on Mars formed following the early Hesperian Period, the youngest of these appearing in the Amazonis and Elysium Planitiae regions just a few tens of millions of years ago. In order for the outflow channels and their associated features to have formed it has been estimated that a volume of water equivalent to a global ocean at least 1 km deep would have been required. As for the discharge rates of the meltwaters, estimates for individual features range between one million and ten billion cubic meters of water per second; the latter is far greater than any known terrestrial floods and is comparable to the flow rates of ocean currents such as the Gulf Stream. Estimated flow speeds are typically between 4 and 60 m/s, but the largest and most catastrophic events may have seen floodwaters exceed velocities of 100 m/s.

Martian valley networks occur across half the planet's surface, mainly through-out the older heavily-cratered southern highlands of Noachian and Hesperian age. Martian valleys can take the form of long, winding valleys with few tributaries, or networks of smaller valleys associated with dendritic tributary patterns. In cases where areas between the individual channels in such features are not cut through by smaller valleys it is thought that they may have formed through sapping, where groundwater seeps onto the surface as a spring and erodes soil downslope.

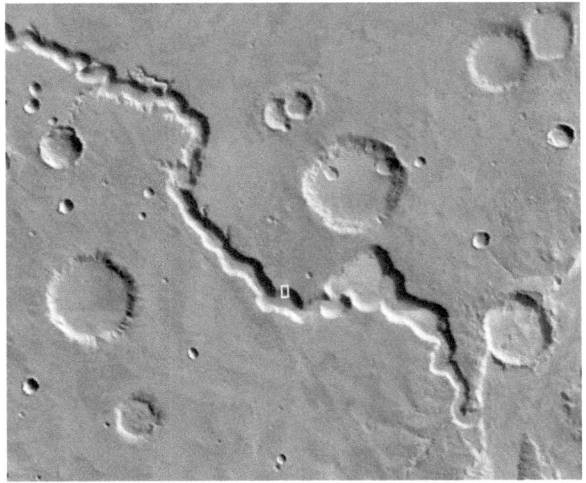

Nirgal Vallis, a narrow meandering river valley (Credit: NASA/JPL)

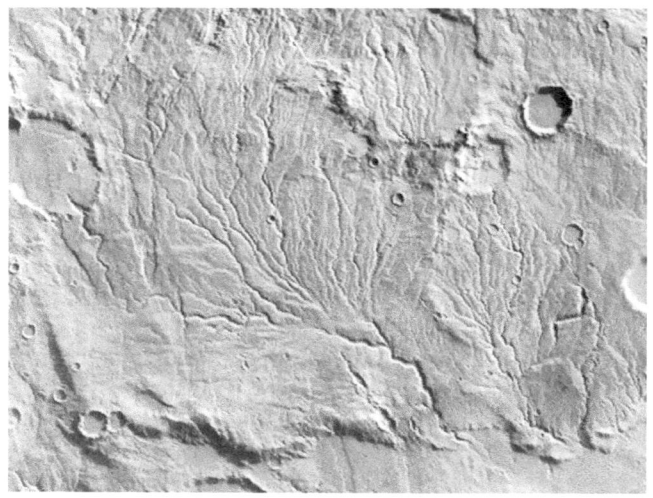

A dendritic valley network common in Mars' southern highlands. This image covers an area about 250 km across (Credit: NASA)

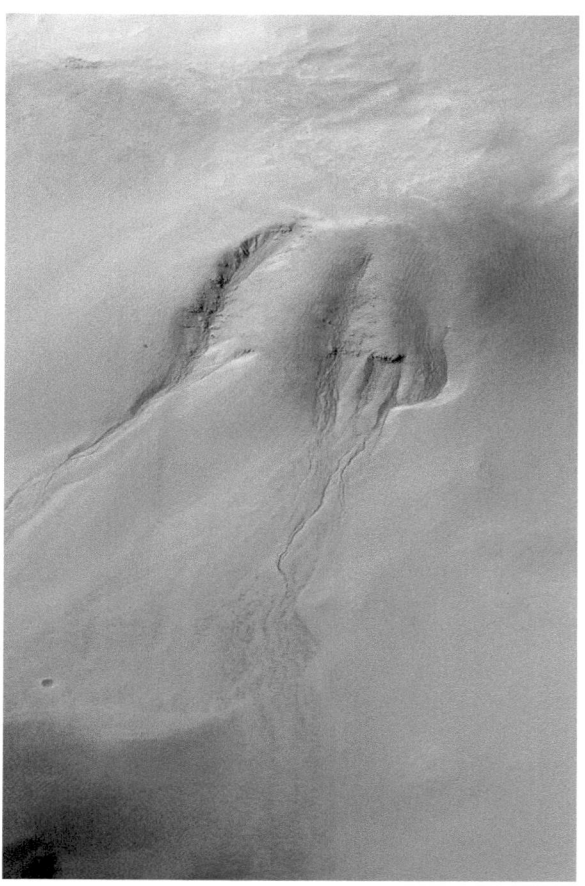

The steep south-facing inner wall of an impact crater in Noachis Terra displays channels formed by sapping. The image covers an area approximately 3 km wide (Credit: NASA)

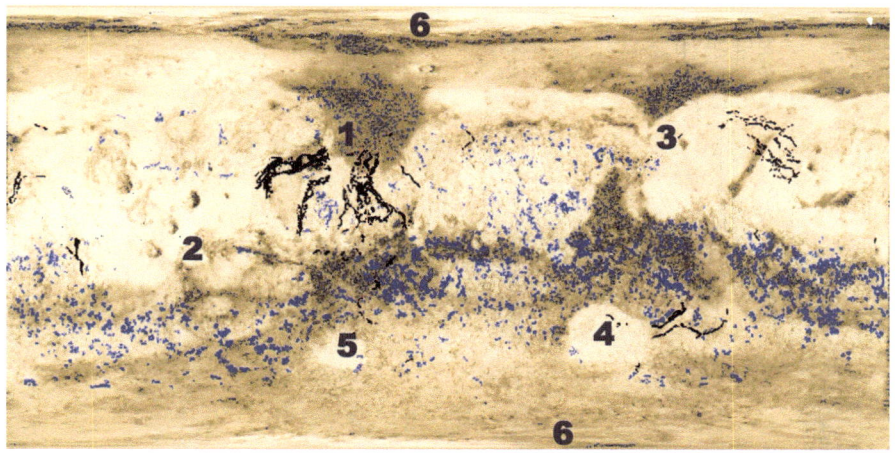

Global map of Martian outflow channels (black) and valley networks (blue). (1) Circum-Chryse region. (2) Tharsis region. (3) Utopia Planitia. (4) Hellas region. (5) Argyre region. (6) Polar regions (Credit: NASA/Grego)

Mars' outflow channels (and features possibly associated with them)

1. **Circum-Chryse region**: Ares Vallis, Ganges Chasma, Kasei Valles, Maja Valles, Mawrth Vallis, Ravi Vallis, Shalbatana Vallis, Simud Valles, Tiu Valles
2. **Tharsis region**: Parts of the Olympica Fossae, valleys adjacent to southeastern Olympus Mons, numerous channels flowing into the Amazonis and Elysium Planitiae, Al-Qahira Vallis, Athabasca Vallis, Grjota Vallis, Ma'adim Vallis, Mangala Vallis, Marte Vallis
3. **Utopia Planitia**: Granicus Valles, Hrad Vallis, Tinjar Valles, Hebrus Valles
4. **Hellas region**: Dao Vallis, Harmakhis Vallis, Niger Vallis
5. **Argyre region**: Uzboi, Ladon, Margaritifer and Ares Valles, Surius, Dzigai, and Palacopus Valles
6. **Polar regions**: Chasma Boreale, Chasma Australe

3.3 Oceanus Borealis?

Might Mars once have had an ocean of liquid water early in its history, when the planet was far warmer and had a thicker atmosphere? Proponents of the hypothesis speculate that around one third of the planet's surface – including the entire low-lying region of the northern hemisphere's Vastitas Borealis basin – was once covered by liquid water to an average depth of 550 m.

Some of this water was lost to the atmosphere by sublimation and then into space by atmospheric sputtering (as atoms in the upper atmosphere are flicked out into space after being hit by energetic particles from space). Some of this water also seeped into the regolith and upper crust to be absorbed into the planet's subsurface cryosphere. As the climate cooled, the remnants of this speculative ocean, named Oceanus Borealis, froze over and was subsequently buried beneath windblown sediments, volcanic ash and (to a far lesser extent) debris ejected from impacts further afield. The ocean's frozen remnants are presumed to lie beneath the plains of Vastitas Borealis.

There appears to be some evidence supporting the past existence of Oceanus Borealis – or, at any rate, evidence showing that there existed extensive lakes of liquid water across northern Mars. The pattern of drainage channels from south to north onto the northern plains, and their morphology, suggests prolonged periods of rainwater feeding rivers that drained into low-lying areas.

Artist's impression of an ancient ocean on Mars, Oceanus Borealis (Credit: Ittiz, public domain)

Topographic features can be seen which bear a striking resemblance to ancient shorelines and river deltas created by sedimentary deposition. One of these ancient shorelines has been identified in Shalbatana Vallis, where a 50 km long water-carved canyon opens up into a wider valley featuring ridges and troughs indicative of beach deposits and a delta around a lake whose area was some 210 km^2 (ten times the surface area of Scotland's Loch Ness) and 460 m deep. It is thought to have formed during the Hesperian Period around 300 Million years following the planet's 'warm and wet' phase; its eventual disappearance, through evaporation and freezing as the climate changed, took place so rapidly that its lowering shoreline had little opportunity to leave behind any further clear topographical traces.

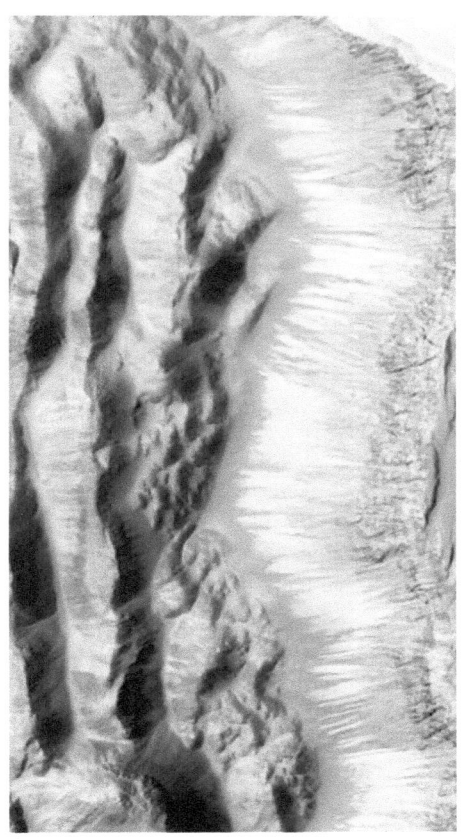

Portion of the floor of Shalbatana Vallis showing an ancient shoreline and delta deposits (Credit: NASA/JPL)

3.4 Liquid Water Today

Large quantities of water ice lie beneath the Martian surface in areas far away from the poles, even along some parts of the equator. Large accumulations of dust hide the ice well from direct imaging, but its presence has been confirmed by instruments on board recent space probes.

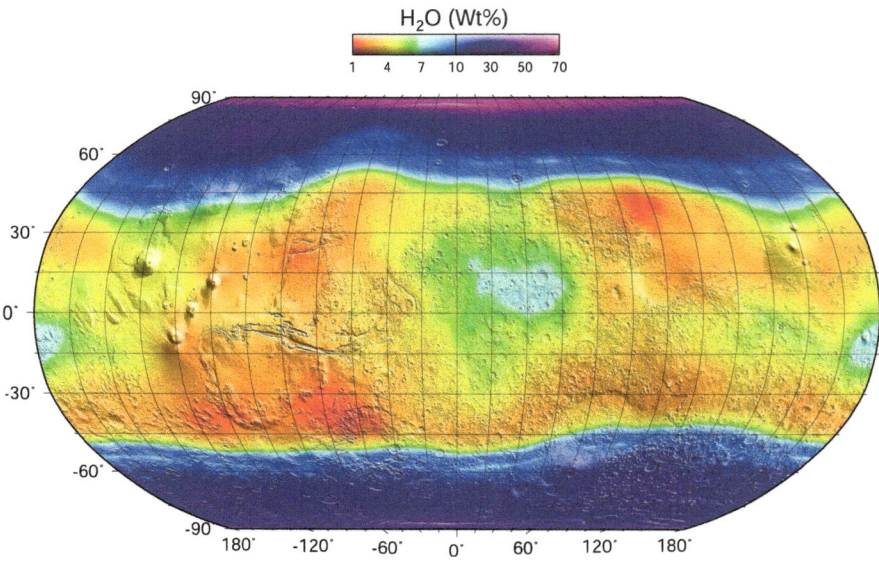

Global map of Martian surface ice (Credit: NASA/Mars Odyssey)

Images from the High Resolution Imaging Science Experiment (HiRISE) aboard the Mars Reconnaissance Orbiter (MRO) revealed evidence for the existence of salty liquid water (brine) on or near the surface of Mars. Thousands of narrow, tendril-like dark streaks have been observed to appear on some of Mars' steep equator-facing slopes during the planet's warmer seasons, when temperatures can rise up to 27°C. Observations show that the streaks can grow by as much as 20 m a day as briny water oozes downhill along narrow, steep channels; the streaks can flow around obstacles and occasionally diverge and converge. As might be expected, the streaks fade or vanish altogether by the time winter sets in and temperatures are too cold even for brine to liquefy. It is thought that salt water in thawing mud is the agent of this phenomenon because it has been observed in conditions too warm for carbon dioxide frost to exist and too cold for the presence of pure liquid water, whereas the conditions are just right for brine with its lower freezing point than water. Intriguingly, these seasonally moistened areas might provide a habitable zone for Martian microbes today. Low levels of exotic gases (such as methane) produced by microbes around these areas might well be detected by sensitive instruments aboard future space probes.

Spring and summer water flows are revealed as dark tendrils in this image (combined orbital image and 3D model) of a slope inside the crater Newton (Image credit: NASA/JPL-Caltech/University of Arizona)

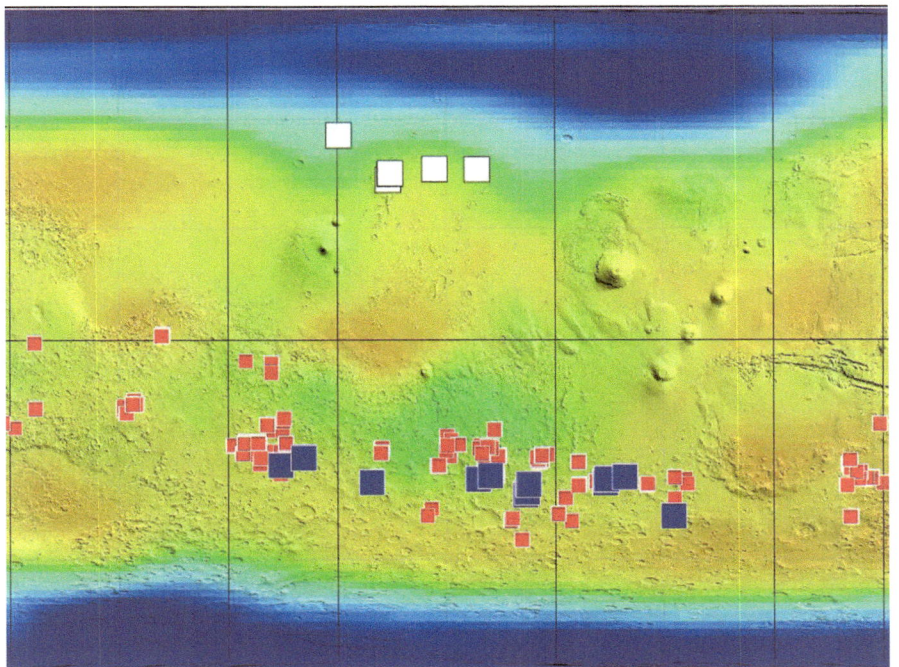

This map of Mars shows the locations of three types of findings related to salt or frozen water, plus a new type possibly related to both salt and water. The map is color coded to concentrations of shallow subsurface water ice. Blue indicates higher concentrations of water ice, orange designates lowest concentrations. White squares mark young impact craters that have exposed water ice close to the surface. Red squares mark possible salt deposits formed by the evaporation of brine. The blue squares mark locations of dark tendrils appearing and incrementally growing down slopes during warm seasons, probably resulting from the action of briny water (Credit: NASA)

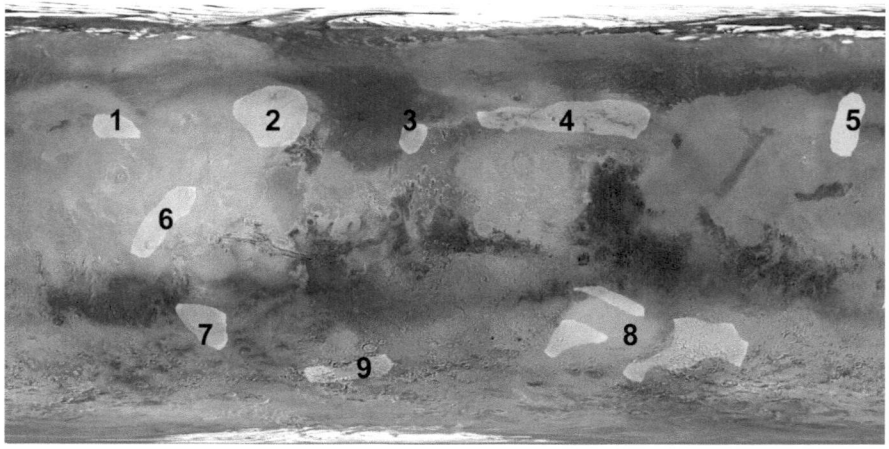

Map of Mars showing areas (brighter patches) of the regolith and upper crust speculated to harbor substantial rock 'glaciers', based upon the observed high density of lobate debris aprons present around young impact craters. These areas appear to run around the planet centered on the 40°N and 40°S latitudes, plus the equatorial region in the vicinity of the Tharsis volcanoes. Key: (1) north of Olympus Mons, (2) Tempe, (3) Acidalia; (4) chaotic terrain around Deuteronilus and Protonilus Mensae, (5) Phlegra, (6) the west escarpment of Olympus Mons, (7) around Claritas Fossae, (8) around Hellas, (9) Argyre (Credit: NASA/Grego)

3.4.1 Polar Ice Caps

Most of the water on Mars is locked up as ice in the thick polar caps and around other parts of the planet in sub-surface 'glaciers'. It is estimated that if all the polar ice melted the planet would be covered by an ocean averaging 11 m deep. Both polar caps are composed of substantial layers of water ice and frozen carbon dioxide (dry ice) and they grow to their largest extent when they are in complete darkness during their respective autumn and winter seasons through an accumulation of carbon dioxide ice, when up to 30% of the carbon dioxide in the Martian atmosphere condenses and freezes.

Mars' eccentric orbit produces an unequal length and severity of winter at the poles. The seasonal north polar cap grows during the 305 day long northern autumn and winter when Mars is near perihelion and extends to a latitude of around 64°N. Mars is near aphelion during the 382 day long southern autumn and winter, when the seasonal south polar cap extends below a latitude of around 55°S.

Cloud and ice-fog hazes form over each pole as the Sun climbs higher during its respective late spring, creating a hazy grey blanket or polar 'hood'. Although warming takes place, the atmospheric pressure is too low to allow ice to melt and become liquid, so the water vapor forming these clouds arises through sublimation. As temperatures rise, strong katabatic winds blow off the poles at speeds of up to 400 km/h. As the cloud and haze dissipates, the brilliant ice caps come into view.

The residual polar caps are on view in these Hubble Space Telescope images showing summer at the northern hemisphere (*left*) and southern hemisphere (*right*). Note the Chasma Boreale in the north polar cap and the 'Mountains of Mitchel' at the edge of the south polar cap (Credit: NASA/STScI)

As the seasonal ice caps of both poles retreat, polar laminated terrain (PLT) is exposed. Formed by winds depositing material on the ice caps, PLT is clearly visible because the layers have varying compositions of dust, sand and volatiles, the layers having been laid down during cyclical variations in the climate of Mars due to the planet's long-term axial 'wobble.' Since ice deposits remain for longer periods towards the poles, PLT layers are generally thicker. Composed of relatively unconsolidated, highly erodible material, wind scouring produces smooth valleys, scarps and spurs; the deposition and removal of PLT does not significantly affect more resilient terrain beneath.

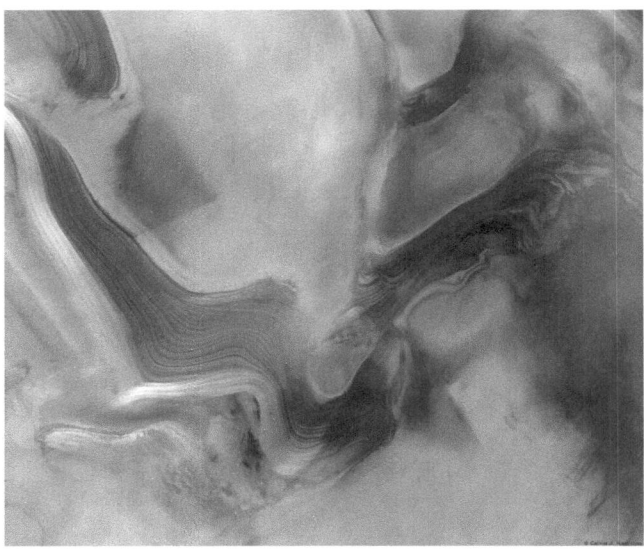

Laminated terrain in Mars' north polar cap (Credit: NASA/JPL)

Close-up an eroded cliff face, revealing layered terrain (Credit: NASA/JPL)

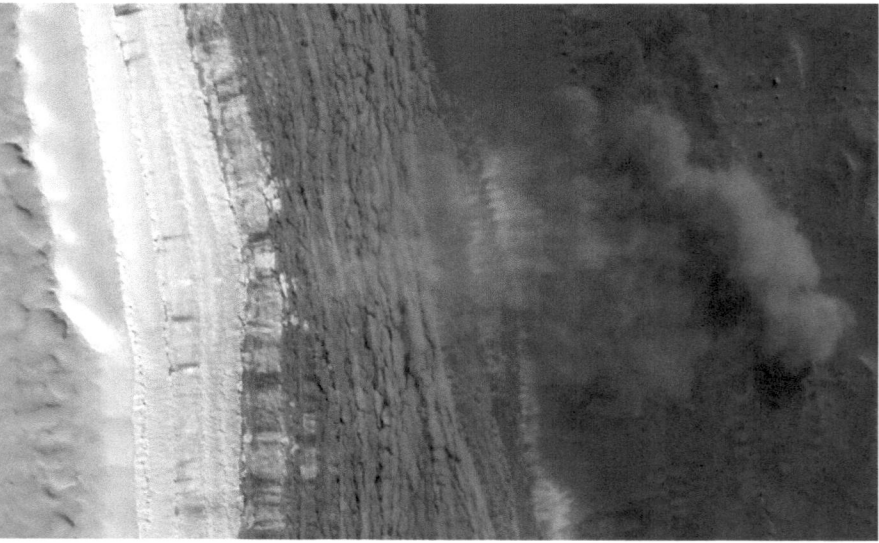

A Martian avalanche caught in action. Dislodged from the steep upper slopes of a 700 m high scarp in the north polar layered deposits (at 83.7°N, 124.2°W), a cloud of fine-grained ice and dust cascades to gentler slopes below; the cloud is about 180 m across and extends about 190 m from the scarp base. It is not known for certain what triggered the event, but spring melting may have caused it (Credit: NASA/JPL)

3.5 North Polar Cap

At the height of northern summer there remains a residual north polar cap composed of water ice. Measuring around 1,100 km across with an area of some 700,000 km^2 and averaging 2 km thick, the residual cap is host to a vast chasm known as Chasma Boreale, in addition to numerous deep grooves that spiral from the cap's center.

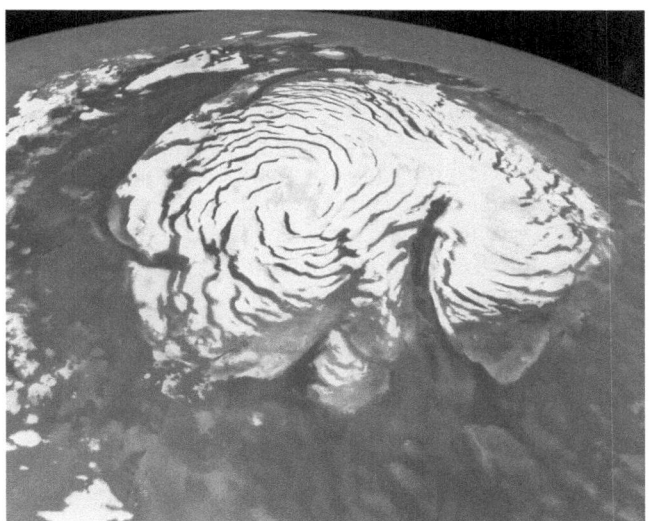

Mars' residual north polar cap. Note the spiral grooves and Chasma Boreale (Credit: NASA)

Almost splitting the residual ice cap into two, Chasma Boreale is a valley 560 km long, 100 km at its widest and with a flat, duned base averaging 1,400 m deep; seen in close-up its walls display layers of material that doubtless hold many secrets about past conditions on Mars. When the chasm was first imaged by space probes most planetary scientists thought that it was likely to have formed by the erosion of layered deposits at the pole – either through a catastrophic flood of water, melting of its base over a long time period, aeolian processes or a combination of these. It is now widely thought that Chasma Boreale formed at the same time that the layered deposits were being laid down, taking on its form through the non-uniform accumulation of these deposits determined by pre-existing topographic features. Indeed, another valley of comparable size and age to Chasma Boreale has been discovered at the north pole, but this one has been in-filled by deposition and is no longer visible as a topographic feature.

A view of Chasma Boreale (Credit: ESA/Mars Express)

Spiral grooves in both the northern and southern ice caps are thought to be formed by high velocity katabatic winds gusting downslope from the poles. Rather than streaking through the grooves and ablating the surface material, the winds blow across the grooves, transporting ice from the upwind to the downwind side, causing the ice on the upwind side to thin and a thickening of the ice on the downwind side of the groove, resulting in a slow migration upwind over time. Mars' rotation produces the spiral shapes of the grooves, as the coriolis force (familiar to us on the Earth in the way that cloud systems move) acts on winds, twisting them to greater extents at progressively lower latitudes. This has produced a strong clockwise swirling of the grooves in the north polar cap and (a less obvious) anticlockwise swirling of those in the south polar cap.

As winter comes to the northern hemisphere, temperatures fall below −123°C and the north polar cap gains a meter-thick coating of frozen carbon dioxide. The cap gradually increases in size, extending down to a latitude of about 55°N.

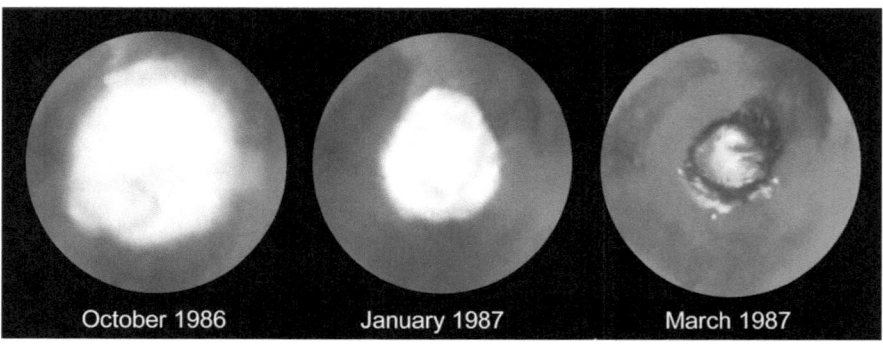

October 1986   January 1987   March 1987

Seasonal changes in Mars' north polar cap, imaged by the Hubble Space Telescope (Credit: NASA/STScI)

3.6 South Polar Cap

Measuring 400 km across and with a thickness of 3 km, the residual south polar cap has two layers; the top layer, around 8 m thick, is made of frozen carbon dioxide, while the lower layer is composed of water ice.

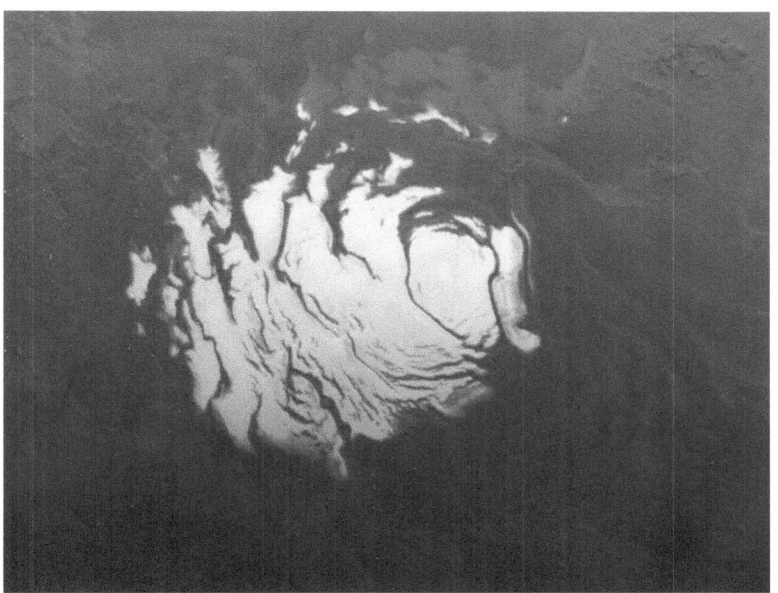

Mars' residual south polar cap (Credit: NASA/Viking)

Surprisingly, the residual cap is not centered on the geographic south pole, but is by offset by 2° in latitude towards Argyre Planitia. This displacement is caused by the circulation of the Martian winds which produces a snow-laden low pressure system in the western hemisphere side of the south pole; since snow reflects sunlight more efficiently than frost, it is cooler and slower to sublimate as spring and summer progress. During winter the south polar cap is symmetrical about the pole; when at its largest it is around 4,000 km in diameter, covering about 20% the entire surface area of Mars.

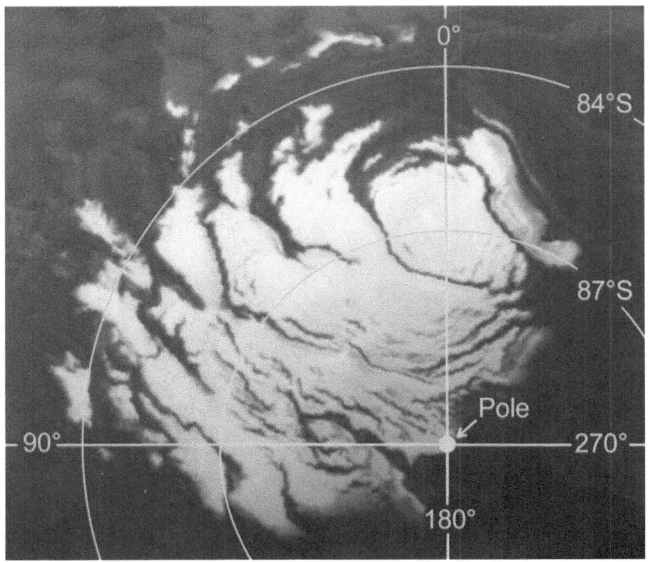

The residual south polar cap is considerably offset from the geographical south pole (Credit: NASA/Grego)

Some 650 km from the south pole (centred at 79°S and 260°W) the 850 km wide plain of Promethei Planum is seasonally blanketed with a deep layer of ice. A 100 km diameter impact crater whose interior is partly covered in ice is visible, as well as structures formed by volcanic lava flows and dark dunes made up of volcanic ash. The south pole lies to the left side of the image (Credit: ESA/Mars Express)

As the seasonal south polar cap retreats during the spring and early summer, its edge begins to show irregularities and a portion of it, south of Hellas Planitia, appears to become isolated from the main body of the cap. Known as the 'Mountains of Mitchel' (not to be confused with Mitchel crater), this is a brighter highland area running around longitude 84°S from 350°W to 40°W. It has a south-facing scarp which, because of its orientation away from the Sun and albedo, hangs on to its carbon dioxide frost coating longer than surrounding areas during the spring.

Early spring over the 'Mountains of Mitchel'. Image covers an area 750 × 300 km (Credit: NASA/MGS)

In contrast with the north polar residual ice cap, whose surface is generally flat and pitted, the south polar residual cap has a complex landscape of broad, flat mesas, small buttes, pits and troughs, in some places resembling stacks of Swiss cheese slices. The scarps and pit walls have been observed to retreat annually (per Martian year) an average of 3 m, and in places up to 8 m, while their floors (which receive less direct sunlight in terms of area) remains the same level. As they retreat, the south polar pits merge and become plains; isolated patches between the plains become mesas which in turn become short-lived buttes.

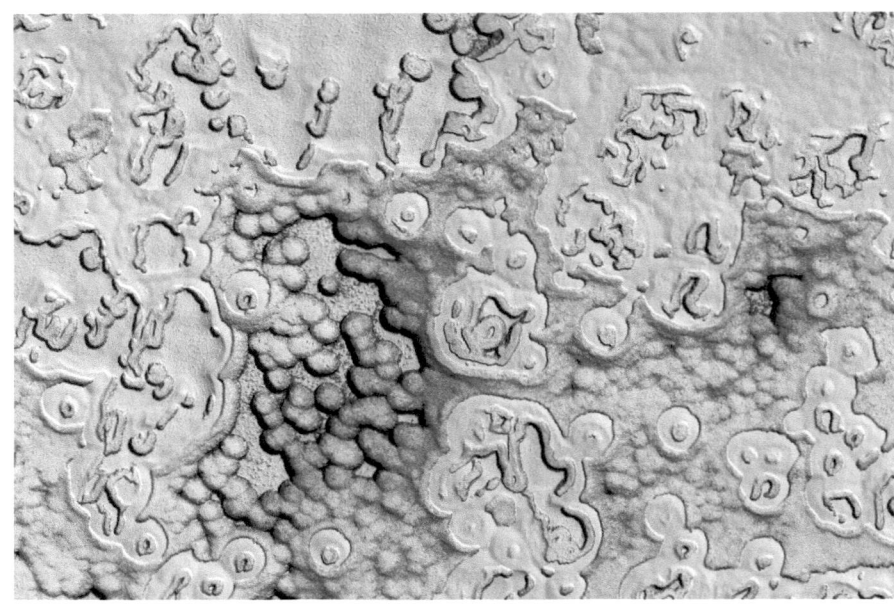

'Swiss cheese' landscape typical of many parts of the south polar residual ice cap's surface. The area imaged measures 7 km across and is centered at 87.6°S, 358.6°W (Credit: NASA/JPL)

Unique to the south polar region in an area between 60 and 80°S and 150 to 310°W are unusual features called 'spiders' (narrow troughs) and bright and dark fans; the area came to be known as 'cryptic terrain' because planetary scientists initially found it difficult to explain their origin. The features are thought to be formed rapidly (over the space of days and weeks) during the spring thaw by small-scale eruptions of carbon dioxide jets, or geysers. As carbon dioxide sublimes beneath meter-thick slabs of transparent carbon dioxide ice at the surface, the gas flows in a radial pattern, producing the 'spider' troughs. In places the gas breaks through to the surface where it escapes in geysers, lofting darker dust particles into the atmosphere; the dust is deposited on the surface of the ice in fan-shaped deposits downwind.

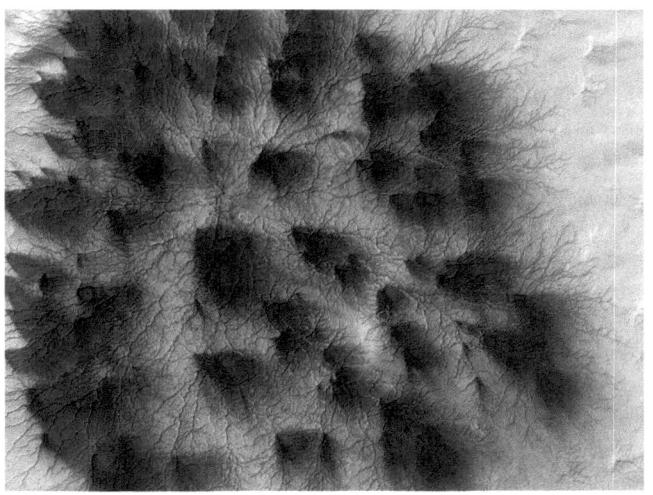

This image, covering an area about a kilometre across (centred at 81.8°S, 76.2°E), shows a network of troughs forming a starburst pattern known as 'spiders' dotted with dark fans (Credit: NASA/JPL)

3.7 Impact Features

Much of Mars has been intensely sculpted by meteoroidal and asteroidal impact. All of the terrestrial planets – Mercury, Venus, the Earth and its Moon – are thought to have been subjected to similar levels of bombardment. Through the aeons, the substantial atmospheres of Venus, the Earth and Mars have served as an effective impact buffer, allowing only the biggest and more substantial incoming objects to wreak large scale damage, while volcanic and tectonic processes went on to completely obliterate the more ancient impact scars. Atmosphere-less Mercury and the Moon have been fully exposed to the harsh vacuum of space and subjected to bombardment by interplanetary dust, meteoroids, asteroids and comets, in addition to X-rays, gamma rays and cosmic rays; their surfaces have been modified to a lesser extent by volcanic and tectonic forces, so many of the ancient impacts are clearly visible.

Mars lies between the two ends of the spectrum, and while many ancient impact basins (particularly in the north) have been obliterated or hidden, some very old impact basins remain clearly visible. Much of Mars' southern hemisphere is older terrain covered with thousands of sizeable impact craters, while most of the surface of the younger northern hemisphere is relatively sparsely cratered. Around 60% of Mars' surface is densely cratered, while the remaining 40% has relatively fewer impact craters. Note that the Borealis Basin, which covers much of the northern hemisphere, may itself have been produced by a single giant impact.

Mars displays around 250,000 craters larger than a kilometer; more than 40,000 are larger than 5 km across and around 150 impact features have diameters in excess of 100 km. Mars' crust has never been subjected to plate tectonics, but large tracts of its surface have been modified by volcanic activity, sedimentary processes, weathering and other geological processes, including metamorphic activity and faulting. Nevertheless, impact features both large and small, formed several billions of years ago, can often be clearly traced.

A relatively small number of Martian craters represent the calderas of ancient volcanoes, but there is no evidence to suggest that any major craters formed by violent crustal explosions or as a result of crustal collapse after magmatic subsidence, for if some craters did happen to have such an origin then a number of these features, frozen in various stages of formation, would likely be observed.

Meteoroidal and asteroidal impact has produced most of Mars' craters, and their appearance has, in various degrees, been modified by volcanic, metamorphic, fluvial and aeolian weathering processes. Numerous older craters have been deformed by tectonic activity and faulting. Relatively younger craters with steep inner slopes often show avalanche streaks, along with deep gullies (some having perhaps formed relatively recently) which appear to have been cut by the erosive action of flowing liquid. Lava flows, sediment from drainage and wind deposition of material has partly in-filled many crater floors. Indeed, some impact features have been almost completely filled-in, the remains of their exposed rims giving rise to 'ghost' craters or quasi-circular depressions.

The overwhelming majority of Martian craters display the hallmarks of impact formation so familiar on all the terrestrial planets and the Moon. There is a clear pattern of Martian impact feature morphology spanning all size ranges, from small meteoritic impact pits to vast asteroidal impact basins. The observed morphology of Martian craters squares with computer studies and ballistic impact experiments performed in laboratories and field studies of terrestrial craters (both natural impact features and manmade explosion craters).

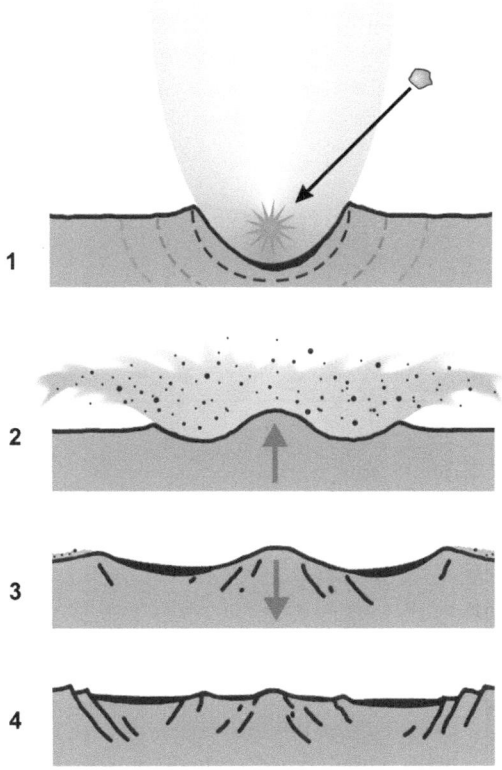

Stages in the formation of a simple impact crater. (1) Impact and excavation. (2) Rebound and uplift, ejecta blanket. (3) Isostatic adjustment, collapse of uplift, wall slumping and faulting, impact melt and possible lava flooding of floor. (4) Final pristine crater (Credit: Peter Grego)

A meteoroid of a size that might burn up completely in the Earth's atmosphere may hit the Martian surface and produce a small impact crater. Craters smaller than about 5 km across are usually smooth, bowl-shaped depressions with a large diameter to depth ratio. Simple ejecta collars – usually an area of rough terrain that extends to about a crater diameter way from the rim – are seen in the fresher examples of such craters, and they can be found all over Mars. Larger impact features generally have a smaller depth to diameter ratio and they display progressively more complex forms.

Depending on the size and nature of the impactor, the impact velocity and the nature of the terrain impacted, the diameter of impact craters on Mars can measure between 10 and 50 times the diameter of the impactor, and the volume of material that they excavate can be hundreds of times the impactor's volume. Impacting at around 6 km/s, the lowest impact velocities are produced by local asteroids on near-circular heliocentric orbits. Other asteroids and short-period comets can impact at twice this velocity, while long-period comets may have impact velocities in excess of 30 km per second.

The depth to diameter ratio of craters on Mars is far more diverse than those of the Moon because of later weathering and infilling by volcanic and/or sedimentary processes. Analysis of pristine Martian craters reveals that the depth-diameter relationship of impact structures ranges from 10:1 in ~10 km craters and 25:1 in the ~50 km crater range to around 100:1 in the larger basins.

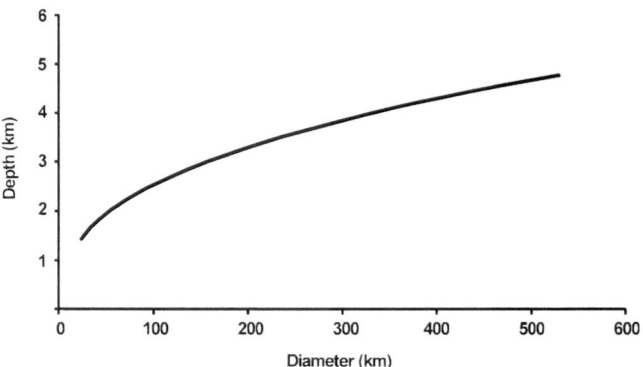

The depth-diameter ratio of large Martian impact features (Credit: Peter Grego)

Depending on a variety of factors (size, composition, impact velocity and angle of approach) some impactors are capable of penetrating the regolith to the solid crust beneath. An impactor substantial enough to slice down into the Martian crust generate tremendous pressures and temperatures as its kinetic energy (the product of the object's mass and the square of its velocity) is converted into shock waves and heat that is imparted into the surrounding crust. Immediately following penetration, the crust beneath the impactor is compressed and shock waves are transmitted through the impactor and the surrounding crust. The surrounding material is pushed downwards and outwards, and jets of material are ejected outwards during the explosion.

An ultra-hot bubble of expanding molten material with a temperature of several million degrees is formed as the impactor and the surrounding rocks are almost instantaneously vaporized. The edge of the forming crater is deformed and uplifted as a plume of excavated material, made up of vaporized rock and larger rock fragments, is blasted outwards from the impact site. As the crust decompresses, rebound effects produce a central uplift in larger craters, and a substantial layer of melted rock accumulates within the crater's bowl. Isostatic adjustment leads to the slumping of inner walls, and the floors of larger craters may fracture, allowing lava to rise onto the surface and flow over the floor.

An ejecta blanket is produced as the excavated material is distributed around the newly-formed crater. Some larger fragments may be hastily welded together in breccias, while smaller droplets of material may cool and solidify in flight, forming tiny glassy beads or larger 'bombs.' In a simple impact on a solid crust, the ejecta is deposited in an ordered manner. Material that was close to the focus of impact near the surface is first to be ejected, high velocity material which is launched steeply above the surface to be deposited at great distances from the crater. As the impact progresses, deeper material is excavated, but as the overall energy of the impact dissipates, progressively slower velocities mean that the ejecta is distributed ever closer to the crater; the deepest excavated bedrock may barely be lobbed over the crater's rim. This produces an inversion of the ejecta blanket's layering, compared with the original layering of the impact site; material once forming the top layers at the impact site are overlain by material that originally lay beneath it.

But the results of impact are often more complicated on Mars than on other terrestrial planets and the Moon. Unusual 'pedestal' craters (averaging ~2–3 km across) found at high northern latitudes are likely to have been formed by impacts which have failed to completely penetrate an ice-rich regolith. The resulting ejecta has a less ice-rich composition which, during warmer periods, protects the underlying ice-rich regolith while the volatiles in the surrounding regolith sublimate; as the surrounding terrain lowers, the crater and its ejecta slowly become elevated above their surroundings, forming a pedestal structure.

Many craters ranging between 5 and 50 km in diameter show lobate ejecta blankets, known as rampart ejecta. Some of these craters are surrounded by peculiar 'pancake' and 'mudsplash' ejecta patterns, indicative of an impact on frozen ground which was heated and fluidized by the heat generated by the impacted. A survey of the locations of such rampart craters suggests the regions in which there exist large quantities of subsurface ice. More than one ejecta lobe is displayed by some craters; such double and multilayered ejecta patterns. Double layered ejecta craters are to be found mainly on the plains of the northern hemisphere, in the Arcadia, Acidalia, and Utopia regions (primarily between 35°N and 65°N), although some have been identified in the southern hemisphere. Of all craters showing ejecta patterns in the northern hemisphere, around 40% of them show double layered ejecta, while in the south the figure is just 10%. In addition, the extent of the outer ejecta layer is usually greater in northern hemisphere craters than those in the southern hemisphere. It is also likely that craters with 'pancake' ejecta, found mainly in the northern plains, represent double-layered ejecta craters whose outer ejecta has been destroyed.

'Pancake' craters (at left) and a multilayer crater (Credit: NASA/JPL/LPI)

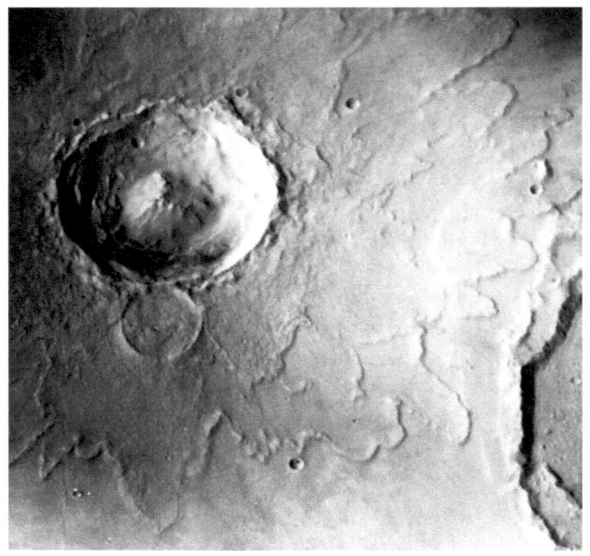

Mudsplash pattern lobate ejecta surround the 20 km diameter crater Yuty, located on the mouth of Ares Vallis in Chryse Planitia (Credit: NASA/JPL/Viking)

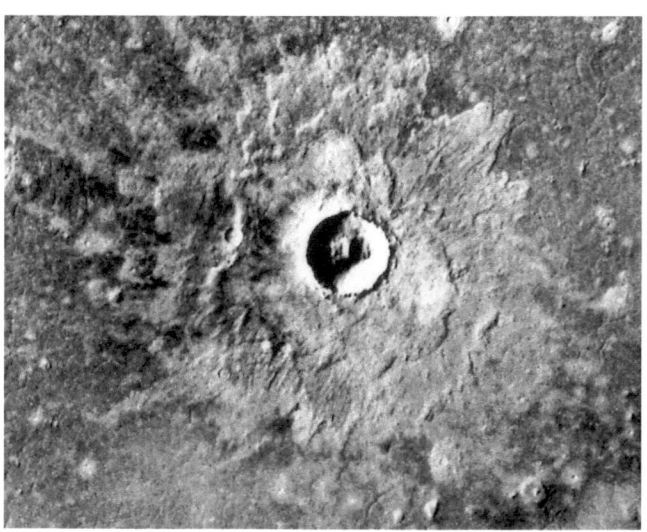

Multilobate ejecta surrounds the 25 km diameter crater Arandas, located in Acidalia Planitia (Credit: NASA/JPL/Viking)

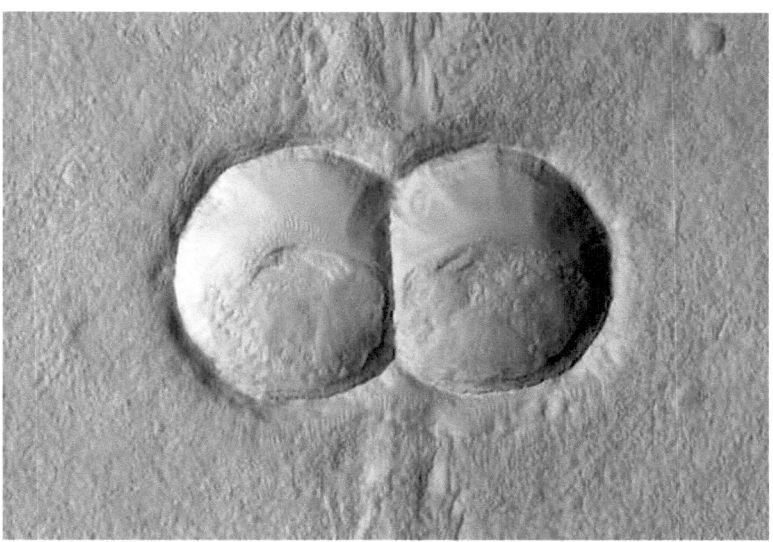

This rare double crater has a shared rim and north-south oriented ejecta deposits. The crater resulted from the simultaneous impact of an object that is likely to have split during its entry into the Martian atmosphere (Credit: NASA/JPL/LPI)

Central pits (both symmetric and asymmetric) are to be found on the floors of many Martian impact craters, some of them at the summits of central elevations. In craters of the 5–125 km size range, nearly 600 centrally pitted craters and more than 330 pitted central elevations have been identified. Although similar-sized craters on the Moon and Mercury sometimes contain such features, they are far more common in Martian craters (and in craters on the ice-rich Jovian moons Ganymede and Callisto) and they are thought to have been formed by impact into a volatile-rich crust, although the precise mechanism for their formation remains uncertain. Martian central pit craters are frequently seen on the rims and in the outer rings of large impact basins, suggesting that the volatiles were originally concentrated in reservoirs created by impact fracturing during basin formation. Around 30% of craters with central pits are found in craters with multilayered ejecta.

Stuff and Substance

A small central pit crater (Credit: NASA/JPL/LPI)

In cases where substantial impacts have excavated material from the crust, beneath a regolith packed with subsurface ice, craters take on an appearance reminiscent of many larger lunar craters. Their interiors are complex, with broad, flattish floors covered with impact melt and occasional lava flooding, at the centres of which rise elevated hills and mountains; terracing is to be found along the inner walls, caused by inward slumping of material. Beyond the rim, their ejecta systems have a less rampart-like, more radial structure consisting of prominent radial ridges and grooves, in addition to secondary cratering caused by the impact of substantial chunks of drier excavated bedrock.

Secondary craters are impact pits formed at a far lower velocity than the original impact. The amount of material excavated by these secondary impacts can actually exceed the volume of material thrown out during the original high-velocity primary impact, since low velocity impactors tend to be more efficient excavators than high energy ones, which generate vast quantities of excess heat. Streams of ejected debris can produce lines of secondary craters arrayed in a radial pattern around their parent crater. These features can consist of an unconnected series of craters of roughly similar size, or in an interconnected chain of craters (some of them highly elongated) that runs a considerable distance across the surface. Such are the dynamics of impact that these radial structures do not always follow perfectly straight lines – secondary chain craters can follow quite curved, even sinuous paths over the surface, and individual craters produced by oblique impacts can take on an elongated appearance.

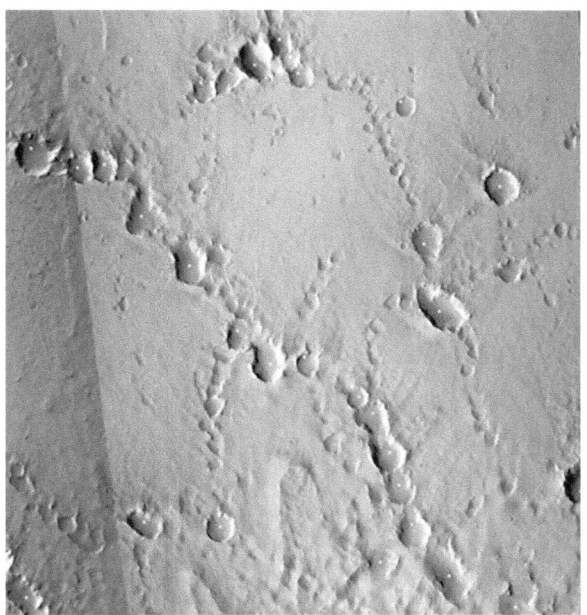

Secondary crater chains, with some components showing elongated outlines. Small white dots indicate secondary craters that are larger than about 1 km in diameter (Credit: NASA/Malin Space Science Systems)

Although most Martian primary craters are roughly circular in outline and have evenly-distributed ejecta systems, this does not necessarily indicate that they were formed by objects falling from anywhere near the vertical. In fact, large impactors hitting the surface at an angle of more than around 12° will produce a circular crater, as they penetrate deep into the regolith and crust before exploding as a point source, spreading ejecta in a pattern all around the crater. At angles shallower than 12° elliptical or elongated primary craters are produced as the explosion is spread out in line with the impactor's original course; often, their ejecta systems are distributed in a 'butterfly' pattern with lobes at right angles to the axis of impact, or spread in a one-sided pattern away from the impact. It is estimated that around 5% of craters larger than 5 km in diameter display elongation, a substantial proportion of which also show butterfly ejecta patterns.

Stuff and Substance

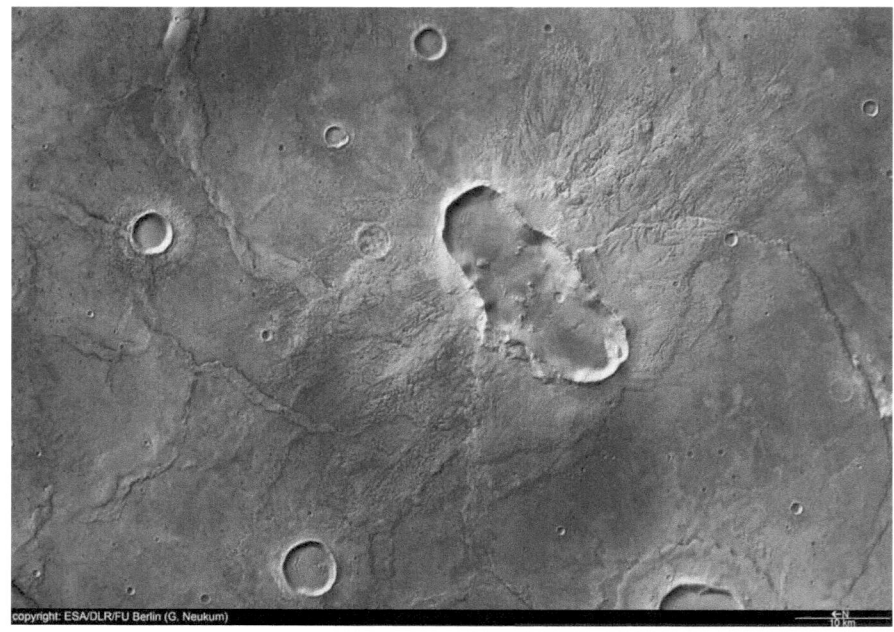

Measuring 24 × 11 km, this sizeable elliptical impact crater lies in Hesperia Planum, its base some 650 m below the surrounding plain. Note the butterfly ejecta system spreading northwest and southeast of the crater (Credit: ESA/Mars Express)

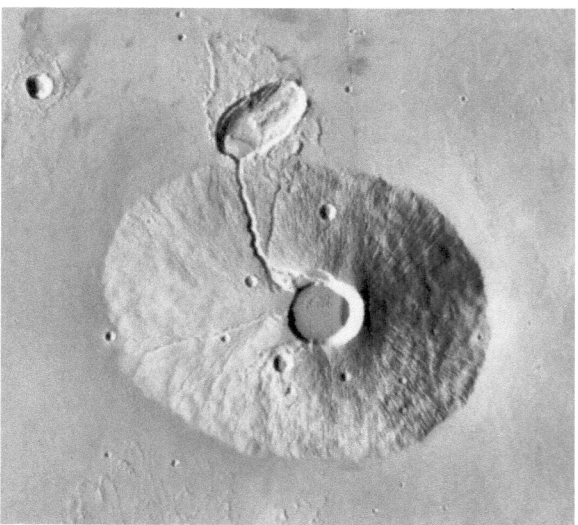

Two elongated craters with different modes of formation. Ceraunius Tholus is a small Tharsis volcano topped by an elliptical caldera; at the northern base of the volcano lies an elliptical impact crater complete with butterfly pattern ejecta. A large sinuous channel joins the two features (Credit: NASA/LPI)

3.8 Ringed Impact Features

Martian craters larger than around 100 km across display elevated rings on their floors rather than central peaks. As their diameter increases increasing diameter, impact features take on the appearance of ripples in a pond spreading in circles from a point; rather than being referred to as craters, the larger impact features are termed ringed basins. The smallest ringed basins often show a substantial central massif, caused by crustal rebound and uplift, surrounded by an inner ring of scattered mountains. Larger basins with diameters between 300 and 1,800 km in diameter have a well-developed inner mountain ring but are lacking in a central elevation, and are similar in appearance to the Orientale basin, the best-preserved ringed basin on the Moon. Larger impact basins between 1,800 and 3,600 km in diameter are referred to as Argyre-type basins, and they show a rugged mountain ring and concentric fault valleys known as grabens.

Lowell, a 203 km diameter basin with an inner mountain ring. It lies in Aonia Terra in Mars' southern hemisphere (Credit: NASA/JPL)

Shaded relief map of Argyre, a multi-ring impact basin 1,800 km in diameter in Mars' southern hemisphere. Argyre Planitia is the smoother central plain (Credit: MOLA Science Team/NASA GSFC Scientific Visualization Studio)

Mars' largest impact structures are multi-ring basins. They display multiple concentric mountain rings which represent frozen waves in the crust produced immediately after impact when the original central uplift collapsed. These rings are riven with faults, scarps and channels, and their outer rings may be far larger than the original asteroid impact excavation; viewed in profile, the largest basins are exceedingly shallow in comparison with their diameters.

Tremendously powerful seismic waves are generated in the crust by a major basin-forming impact, and it has been demonstrated that the focusing of such waves as they converge at a point antipodal to the site of impact led to the formation of chaotic terrain on both Mercury and the Moon. However, unlike Mercury and the Moon, much of Mars' surface has been extensively modified since the late heavy bombardment, so evidence for seismic wave focusing is by no means clearly defined. It has been suggested, for example, that the Hellas impact caused fracturing of the crust at the antipodeal point, which led to volcanism and the formation of the Tharsis volcano Alba Mons.

Martian impact basins and craters larger than 150 km

Name	Diameter (km)	Centre
Borealis basin (speculative)	10,600	Northern hemisphere
Utopia basin	3,200	25.3°N, 212.8°W
Hellas basin	2,300	42.7°S, 290.0°W
Chryse basin	1,600	26.7°N, 40.0°W
Isidis basin	1,500	12.9°N, 270.0°W
Argyre basin	1,100	49.7°S, 56.0°W
Huygens	467	14.0°S, 304.4°W
Schiaparelli	458	2.8°S, 343.2°W
Cassini	408	23.4°N, 327.9°W
Antoniadi	394	21.3°N, 299.2°W
Tikhonravov	386	13.5°N, 324.2°W
Koval'sky	309	30.2°S, 141.5°W
Herschel	305	14.7°S, 230.3°W
Newton	298	40.8°S, 158.1°W
Copernicus	294	49.2°S, 169.2°W
de Vaucouleurs	293	13.5°S, 189.1°W
Schroeter	292	1.9°S, 304.4°W
Newcomb	252	24.4°S, 359.0°W
Flaugergues	245	17.0°S, 340.8°W
Lyot	236	50.8°N, 330.7°W
Secchi	234	58.3°S, 258.1°W
Kepler	233	47.1°S, 219.1°W
Galle	230	51.2°S, 30.9°W
Vinogradov	224	20.2°S, 37.7°W
Schmidt	213	72.3°S, 78.1°W
Mutch	211	0.6°N, 55.3°W
Kaiser	207	46.6°S, 340.9°W
Lowell	203	52.3°S, 81.4°W
Schöner	195	20.1°N, 309.5°W
Dawes	191	9.3°S, 322.0°W
Phillips	190	66.7°S, 45.1°W
Savich	188	27.8°S, 264.0°W
Ptolemaeus	185	46.2°S, 157.6°W
Green	184	52.7°S, 8.4°W
Molesworth	181	27.4°S, 210.9°W
Baldet	180	22.8°N, 294.6°W
Darwin	178	57.3°S, 19.5°W
Terby	174	28.3°S, 285.9°W
Flammarion	173	25.4°N, 311.8°W
Wallace	173	52.9°S, 249.4°W
Becquerel	171	22.3°N, 8.0°W
Henry	171	10.9°N, 336.7°W
Stoney	171	69.8°S, 138.6°W
Mariner	170	35.1°S, 164.5°W
Proctor	168	48.0°S, 330.5°W
Gusev	166	14.7°S, 184.6°W
Denning	165	17.7°S, 326.6°W
Bakhuysen	161	23.3°S, 344.4°W
Miyamoto	160	2.9°S, 7.0°W
Schaeberle	160	24.7°S, 309.9°W
Graff	158	21.4°S, 206.3°W
Janssen	158	2.7°N, 322.5°W

(continued)

(continued)

Name	Diameter (km)	Centre
Dejnev	156	25.5°S, 164.8°W
Lohse	156	43.7°S, 16.8°W
Gale	155	5.5°S, 222.3°W
Trouvelot	155	16.2°N, 13.1°W
Arago	154	10.2°N, 330.2°W
Holden	154	26.4°S, 34.0°W

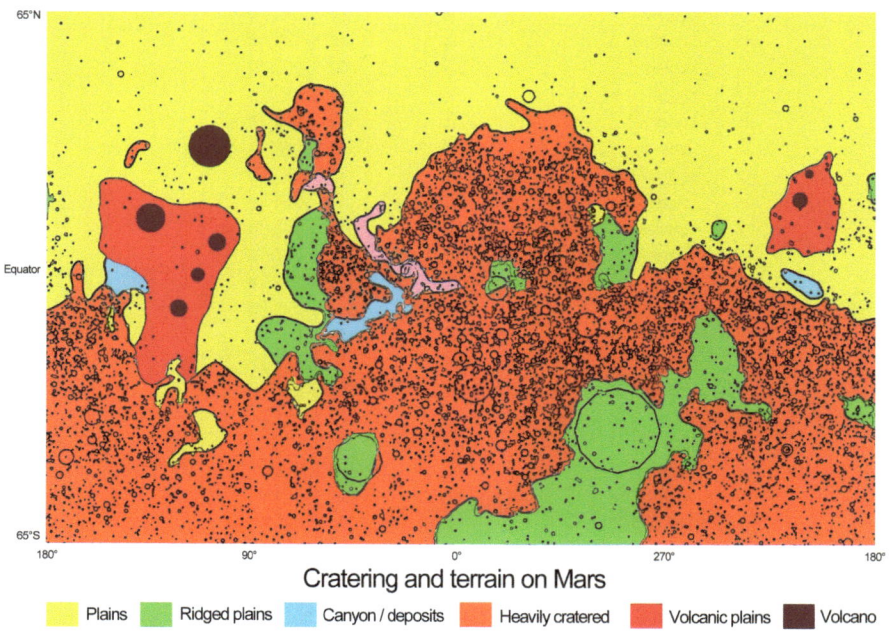

Cratering and terrain on Mars

Plains Ridged plains Canyon / deposits Heavily cratered Volcanic plains Volcano

The distribution of all unburied Martian craters larger than 15 km across (between ± 65° latitude) (Credit: LPI)

3.9 Volcanic Features

In common with all the Solar System's terrestrial planets, Mars has experienced extensive volcanic activity, the products of which are clearly visible in the planet's topography.

Earth-type mantle convection currents never really became established within Mars, so the crust-shuffling process of plate tectonics had little opportunity to get started. Mars' crustal magnetism shows large-scale striping, where extruded magma has been magnetized to the planet's magnetic field. Magnetic striping is familiar to terrestrial geophysicists; it occurs along the mid-oceanic ridges where the crust is being pushed apart on either side, the fresh intrusive material being magnetized according to the magnetic conditions present as the rock cools. As the

Earth's global magnetic field switches polarity (as it does at intervals between a few tens to several millions of years) the changes are recorded in the rocks, preserving an important record millions of years old.

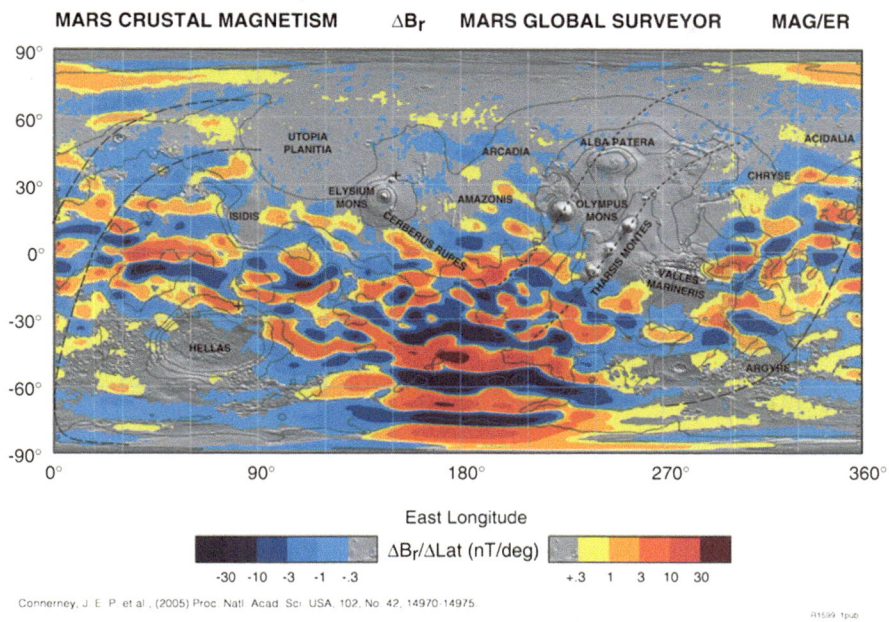

Map of Mars, showing stripes of crustal magnetism. Magnetic north is red, blue is magnetic south (Credit: NASA/MGS)

Extensive basaltic marial plains can be found east of the Tharsis uplift, south of Hellas, in Hesperia Planum and Syrtis Major Planum. These plains date from the same time as the lunar maria, erupting around 3.5 to 3 billion years ago. We only see part of the picture of Martian vulcanism because much has been modified, eroded or buried by sediment since this distant epoch.

Some of the smaller shields and volcanic cones on Mars were active two billion years ago, while the giant shield volcanoes formed between two and one billion years ago. It is unlikely that Mars is volcanically active; the final stages of Martian vulcanism are thought to have taken place between 200 and 20 million years ago, when lava flows rolled down the slopes of Olympus Mons.

Many volcanic features familiar to us on the Earth, produced by plate tectonics, are absent on Mars. For example, Mars has no 'rings of fire' caused by crustal sub-duction, nor has it anything corresponding with our mid-ocean volcanic ridges. Given the absence of plate tectonics on Mars, long-lived mantle upwelling produced large magma chambers deep beneath the surface which caused the crust above to uplift; eruptions in these uplifted areas often were long-lived and effusive, causing the growth of several very large static volcanic shields, similar to the volcanic shields of the Hawaiian chain. Although Mars has a surface area of just 38% that of the Earth, in scale terms its shield volcanoes far exceed the dimensions of those found on our own planet, with more extensive lava flows – a product of the planet's lower gravity and longer-lived eruptions.

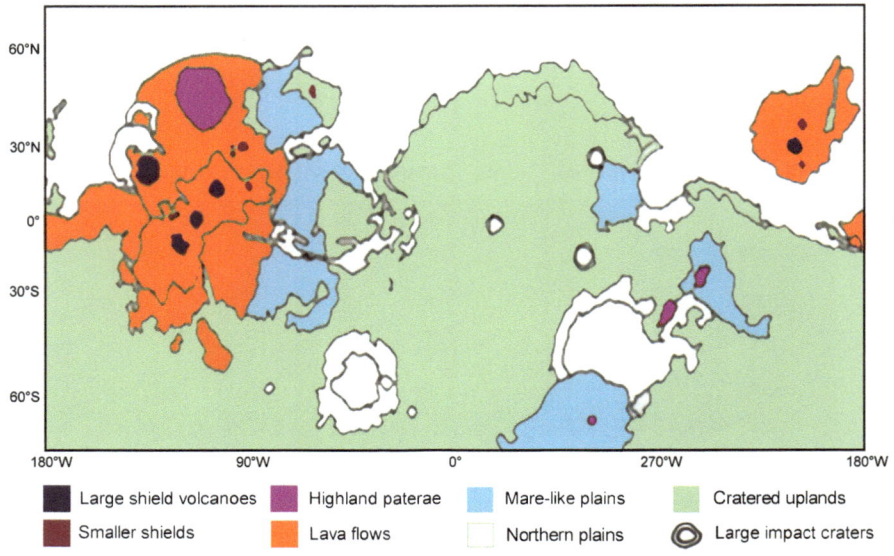

| Large shield volcanoes | Highland paterae | Mare-like plains | Cratered uplands |
| Smaller shields | Lava flows | Northern plains | Large impact craters |

Map of Mars, showing volcanic areas (Credit: NASA/Grego)

Four immense shield volcanoes dominate the vast uplifted Tharsis region, namely Arsia Mons, Pavonis Mons, Ascraeus Mons and the mighty Olympus Mons. These shields were active for many hundreds of thousands of years. Typically, the shields have slopes of around 6° – hardly precipitous, despite their impressive appearance in space probe images. Some 110° west of Tharsis is the smaller uplifted 'island' of the Elysium region whose largest volcano, the shield of Elysium Mons, is flanked in the north by Hecates Tholus and in the south by Albor Tholus. Each volcano is capped by a large summit crater or crater group, known as a caldera, formed as the summit collapsed after volcanic activity ended.

Shield volcanoes formed by quiescent, effusive eruptions are the largest and most striking type of Martian volcano, but there are other types to be found. Many smaller volcanoes formed as a result of explosive eruptions, built up of successive deposits of volcanic ash and viscous lava; as a result, these features have a more convex appearance and have steeper slopes than the shields. Ceraunius Tholus, Tyrrhenus Mons and Hadriaca Patera are examples of such volcanoes; ash clouds were dispersed widely from powerful Plinian-type eruptions and the prevailing winds generally deposited this material to the east of the volcano, covering the terrain in layers with each successive eruption, sometimes interspersed nearer the vent with effusions of viscous lava.

Ash deposits are more friable than the underlying landscape, so they succumb more easily to erosional forces; characteristic of these ash deposits is ridged and trenched terrain, found in areas downwind of many Martian volcanoes, caused by Aeolian erosion. Widespread ash deposits from the high-altitude calderas of shield volcanoes are also in evidence. For example, ash from Apollinaris Patera is thought to be the main substance cut through by the aeolian erosional features of Medusae Fossae. However, some ash deposits have no obvious volcanic source; for example, the fine-grained layered deposits in the vicinity of Meridiani Planum cannot be reconciled with any volcanic feature in the area, so their existence remains to be explained.

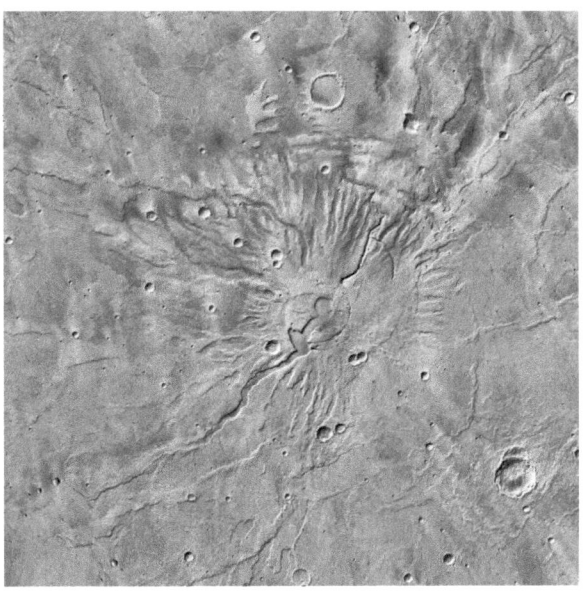

Tyrrhena Patera, an ancient low-profile volcano to the northeast of Hellas Planitia in Mars' southern hemisphere. Its summit, surrounded by arcuate graben, is crowned by an irregular depression that extends down the outer flanks along a valley, perhaps a collapsed lava tube – a similar valley runs down the opposite slope. Although Tyrrhena Patera is considerably degraded, its low profile and the presence of mesa-like formations in its vicinity, suggests that the area is covered with ash deposits rather than lavas. The image shows an area around 250 km wide (Credit: NASA/JPL)

Volcanic activity dwindled as the magma chambers feeding the activity cooled and withdrew deeper beneath the surface, and in some instances large sections of the shields collapsed as the material from which they were composed was too weak to remain intact. Known as paterae (singular, patera, from the Latin for a saucer-like pottery or metal vessel), these collapse features have shallow outer slopes, some with scalloped edges formed by successive collapses. Their volcanic nature is evident in cases where they lie atop an elevation; however, some paterae have been surrounded or embayed by volcanic material, rendering their original volcanic glacis indistinguishable.

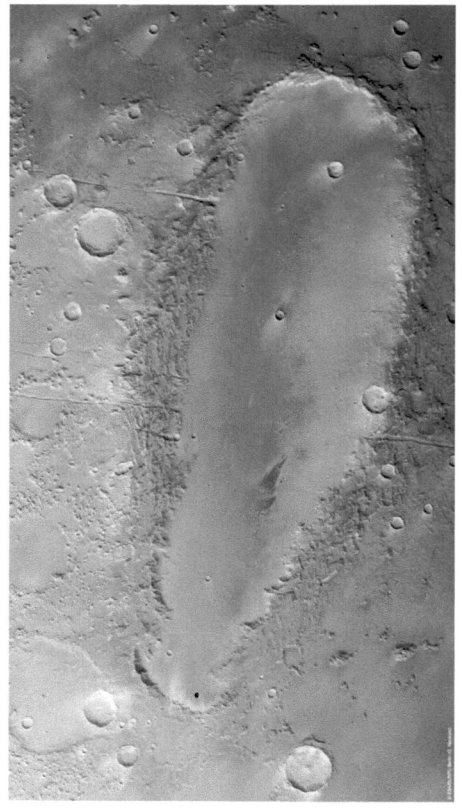

Orcus Patera, a large elliptical depression measuring 380 × 140 km, located just north of the equator at the border of Elysium and Amazonis Planitia. It is likely to be a collapsed volcanic feature, but there is a possibility that it is a large elliptical impact crater (Credit: ESA/Mars Express)

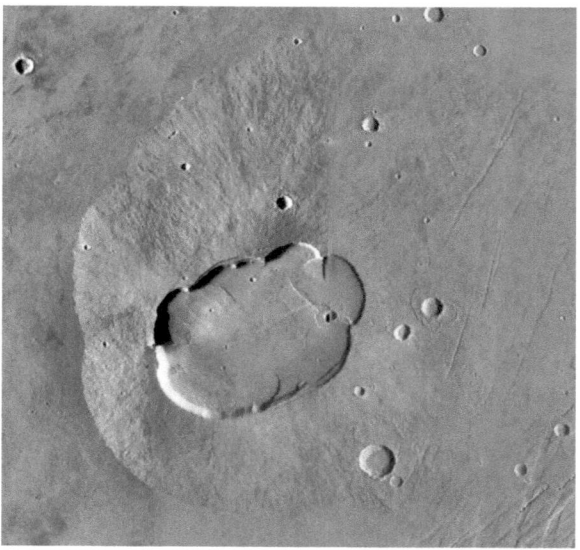

Uranius Patera, an elongated 'crater' atop one of the smaller Tharsis volcanoes. 100 km in length, the feature is a caldera comprising half a dozen circular depressions formed by subsidence when the lava dwindled (Credit: NASA/LPI)

Named mountains on Mars

Name	Centre	Diameter (km)	Elevation[a]	Type[b]
Alba Mons	40.5°N, 109.6°W	530	6,800	V
Albor Tholus	19.0°N, 209.6°W	170	4,500	V
Anseris Mons	30.1°S, 273.4°W	58	4,200	M
Apollinaris Tholus	17.9°S, 184.3°W	35	3,200	V
Arsia Mons	8.4°S, 121.1°W	475	17,800	V
Ascraeus Mons	11.9°N, 104.5°W	460	18,200	V
Ausonia Montes	27.7°S, 261.2°W	158	1,370	M
Australe Montes	80.3°S, 345.9°W	387	5,000	M
Centauri Montes	38.9°S, 264.8°W	270	1,400	M
Ceraunius Tholus	24.0°N, 97.4°W	130	8,500	V
Chalce Montes	54.0°S, 37.9°W	95	2,300	M
Charitum Montes	58.3°S, 40.2°W	850	2,500	M
Coronae Montes	34.9°S, 273.6°W	236	−2,600	M
E. Mareotis Tholus	36.2°N, 85.3°W	5	640	V
Echus Montes	8.2°N, 78.0°W	395	−90	M
Elysium Mons	25.3°N, 212.8°W	401	13,860	V
Erebus Montes	36.0°N, 175.0°W	785	−3,100	M
Euripus Mons	45.1°S, 255.0°W	91	4,480	M
Galaxius Mons	35.1°N, 217.8°W	22	−3,900	M
Geryon Montes	7.8°S, 82.0°W	359	2,250	M
Gonnus Mons	41.6°N, 91.0°W	57	2,890	M
Hecates Tholus	32.4°N, 209.8°W	183	4,720	V
Hellas Montes	37.9°S, 262.3°W	153	1,310	M
Hellespontus Montes	44.7°S, 317.2°W	730	−1,370	M
Hibes Montes	3.7°N, 188.7°W	137	−1,460	M
Horarum Mons	51.4°S, 36.6°W	20	−760	M
Issedon Tholus	36.3°N, 95.0°W	52	830	V
Jovis Tholus	18.4°N, 117.5°W	58	2,990	V
Labeatis Mons	37.8°N, 76.2°W	23	1,860	V
Libya Montes	2.8°N, 271.1°W	1,170	2,100	M
N. Mareotis Tholus	36.7°N, 86.3°W	3	820	V
Nereidum Montes	38.9°S, 44.0°W	1,130	1,920	M
Oceanidum Mons	55.2°S, 41.3°W	33	−1,790	M
Octantis Mons	55.6°S, 42.9°W	18	−1,490	M
Olympus Mons	18.4°N, 134.0°W	648	21,170	V
Pavonis Mons	0.8°N, 113.4°W	375	14,030	V
Peraea Mons	31.4°S, 274.0°W	22	−840	M
Phlegra Montes	41.1°N, 194.8°W	1,352	−1,170	M
Pindus Mons	39.8°N, 88.7°W	17	1,140	M
Scandia Tholi	74.0°N, 162.0°W	480	−4,800	V
Sisyphi Montes	69.9°S, 346.1°W	200	1,370	M
Sisyphi Tholus	75.7°S, 18.5°W	25	2,100	M
Syria Mons	13.9°S, 104.3°W	80	6,710	V
Tanaica Montes	39.8°N, 91.1°W	177	1,980	M
Tartarus Montes	16.0°N, 193.0°W	1,070	−940	M
Tharsis Montes	1.2°N, 112.5°W	1,840	18,200	V
Tharsis Tholus	13.5°N, 90.8°W	158	8,930	V
Tyrrhenus Mons	21.4°S, 253.6°W	473	2,930	V
Uranius Mons	26.8°N, 92.2°W	274	4,853	V

(continued)

Stuff and Substance

(continued)

Name	Centre	Diameter (km)	Elevation[a]	Type[b]
Uranius Tholus	26.1°N, 97.7°W	62	4,290	V
W. Mareotis Tholus	35.8°N, 88.1°W	12	1,250	M
Xanthe Montes	18.4°N, 54.5°W	500	−1,620	M
Zephyria Tholus	20.0°S, 187.2°W	31	2,830	V

Mons. A large, isolated, mountain

Montes. A mountain range (plural of mons)

Tholus. A small, squat, dome-shaped mountain

Tholi. A group of small mountains (plural of tholus)

[a]Elevation. Highest point of mountain or range, measured with reference to datum (mean Martian surface level), not measured relative to the feature's immediate surroundings

[b]Type. V indicates volcanic feature, M indicates non-volcanic mountain(s)

Recent high-resolution images from space probes have revealed the existence of a number of cave entrances at the Martian surface, thought to be vertical shafts produced by the local collapse of layers of ash deposits and lava flows on the flanks of volcanoes. One of the largest 'hauls' of caves discovered on Mars – no fewer than seven of them, collectively known as the 'Seven Sisters' – was identified on the flanks of Arsia Mons. With entrances measuring between 100 and 250 m in width, the way in which they are illuminated during the Martian day (sunlight fails to light up most of their floors) suggests that they are at least 73 to 96 m deep, but likely to be much deeper and widening below the surface if they happen to extend along hidden lava tubes.

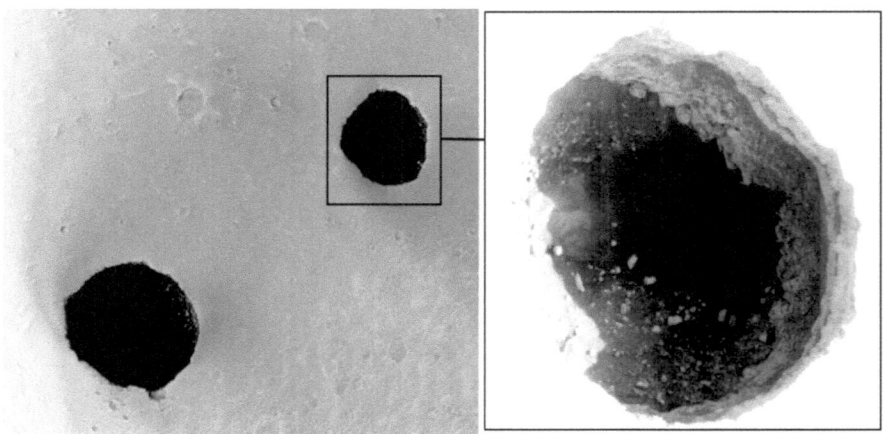

Caves on the dusty flanks of Ascraeus Mons; the overexposed image at right shows some of the poorly-illuminated internal detail which includes soil and boulders, but the end of the cavern cannot be seen (Credit: NASA/JPL)

3.10 Rift Valleys

Although Mars' crust has not succumbed to Earth-like plate tectonics, there are areas where immense crustal tensions have exceeded the yield strength of the rock, causing faulting and extensive rifting. Radiating around Tharsis, dozens of large linear rift valleys can be found, having been caused when the Tharsis region uplifted and bulged, causing tensions which pulled apart the crust. Also in evidence are cross-cutting rift valleys showing that the crustal stresses came from various directions over time. Crustal tension is likely to have arisen from uplifting of the crust due to magmatic intrusion and volcanic loading where the crust is deformed by the weight of volcanic material accumulating on top of it.

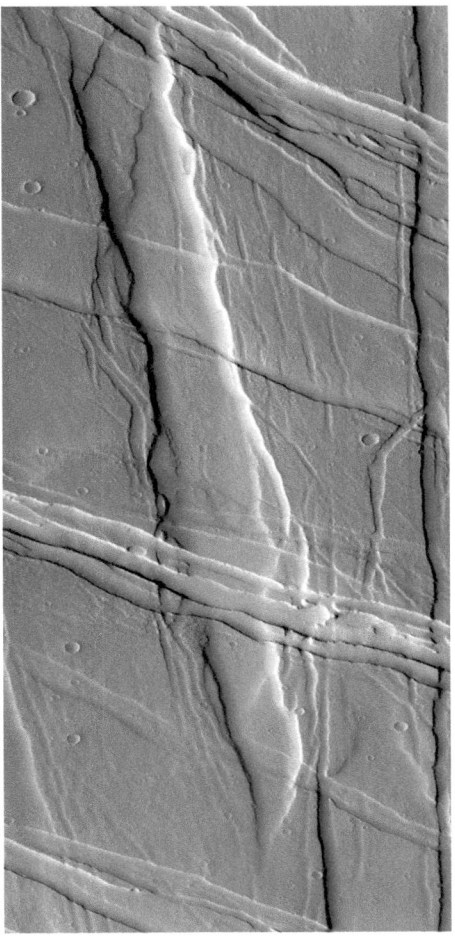

Ulysses Fossae, in the Tharsis volcanic region, shows numerous cross-cutting rift valleys formed by tectonic processes as the area experienced tensional stresses in various directions (Credit: NASA/JPL)

An extensive mass of roughly parallel north-south rift valleys, the Ceraunius Fossae, is to be found in the vicinity of Alba Mons, an immense 530 km diameter shield volcano on the northern side of the Tharsis plateau. Along some of these linear valleys, small collapse craters appear to have been formed, many appearing as isolated pits or, where a number of pits have coalesced, chain craters. Pit and chain craters formed by collapse can usually be distinguished from impact features because they lack raised rims and surrounding ejecta deposits.

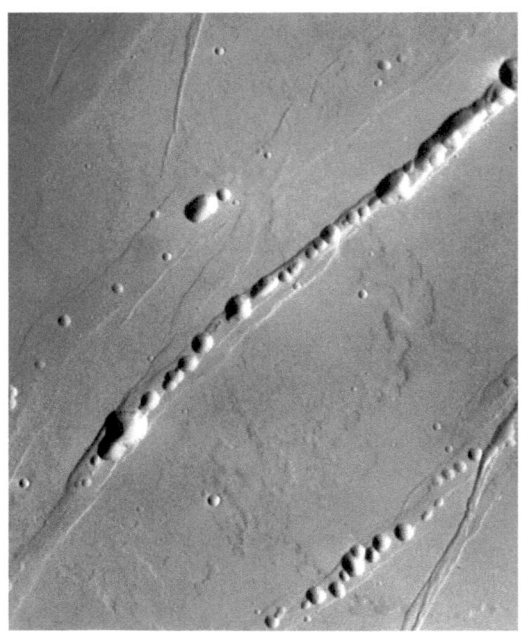

Close-up of an area of the northern Ceraunius Fossae, where collapse pits and chains can clearly be seen lying along rift valleys (Credit: NASA/JPL)

A vast, interconnected system of broad, deep valleys, known collectively as Valles Marineris (named after the Mariner 9 Mars orbiter which imaged it in 1971) cuts into the southeastern side of Tharsis. Almost 5,000 km long from east to west, and in places up to 8 km deep, the Marineris system extends almost a quarter of the way around the planet, just south of the equator. The western end of Valles Marineris is a chaotic hive of canyons and gorges known as Noctis Labyrinthus. Dozens of smaller canyons (most of which would dwarf the Grand Canyon in the United States) reach into the sides of the main valley; some have been formed by rifting and subsidence, others by fluvial erosion. At the eastern end of Valles Marineris, the deep valley Coprates Chasma feeds into Eos Chasma, which broadens out into an interconnected series of depressed knobbly plains and valleys; some of these valleys appear to have been altered by flowing water, with streamlined terrain around features of pronounced relief. It is an amazing terrain, one of the visual wonders of the Solar System.

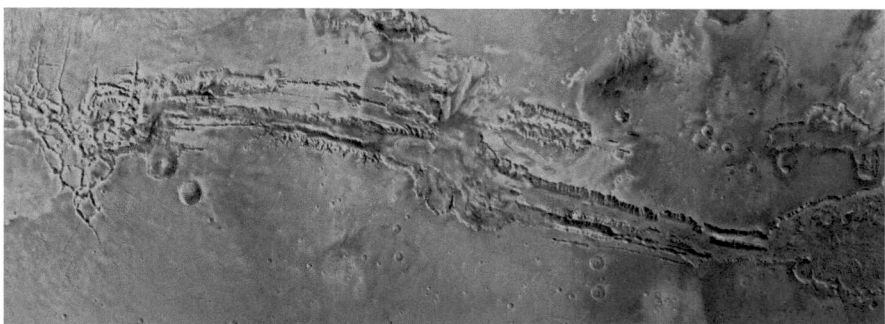

The vast rift valley system of Valles Marineris (Credit: NASA/JPL)

Chapter 4

Atmospherics, Meteors, and Magnetic Field

Even early seventeenth century telescopic observers suspected that Mars has an atmosphere – but then again, the same observers thought that such a commodity was possessed by the Moon, even though there was little evidence to support the idea. During the late eighteenth century William Herschel (1738–1822) – a great proponent of an intelligently inhabited Mars – observed the planet's changing seasons, discerned seasonal variation in the extent of the Martian polar caps and noted apparent variations in the planet's markings which to him suggested occasional cloud cover. In order to get a good view of Mars' surface features terrestrial observers require the weather to be good on the Earth, as well as on Mars!

Although the surface gravity of Mars is only a little more than one-third that of the Earth, the lower Martian temperatures and higher average molecular weight of the atmosphere allows the planet to retain a considerable atmosphere with a scale height (the vertical distance over which atmospheric pressure changes logarithmically) of about 10.8 km, almost 5 km higher than that of the Earth.

Mars' tenuous atmosphere consists of 95% carbon dioxide, 3% nitrogen and 1.6% argon, with traces of oxygen, water and methane. All the gases making up the Martian atmosphere can also be found in the air that we breathe on the Earth – but in completely different proportions. Most of the Earth's air is composed of nitrogen. Carbon dioxide, the most common gas in Mars' atmosphere, is only a minor constituent of the Earth's atmosphere, while oxygen, a gas so vital to life on Earth, is present only as a minor constituent of the Martian atmosphere. An astronaut breathing in a small sample of Martian air wouldn't come to much harm, but it contains far too little oxygen for human survival.

4.1 Methane Mystery

Intriguingly, methane gas has been detected in considerable quantities in the Martian atmosphere, in the order of some 30 parts per billion by volume. Initially suspected in spectroscopic observations from Earth, methane was first definitively detected from orbit in 2003. Mars' methane emanates on a seasonal basis during the warmer northern spring and summer from specific regions and is distributed in large plumes which dissipate within a year. Methane plumes have been mapped

P. Grego, *Mars and How to Observe It*, Astronomers' Observing Guides,
DOI 10.1007/978-1-4614-2302-7_4, © Springer Science+Business Media New York 2012

over the volcanic regions of Tharsis, Elysium and Syrtis Major, in addition to Arabia Terra, a region with large quantities of subsurface water ice; one particular hotspot, Nili Fossae, is an area of deep crustal fissuring around the Isidis impact basin and is covered with hydrated mineral clays. Observations during the Martian summer found that the rate of methane release from the largest plume was equivalent to that emanating from the natural marine hydrocarbon seepage field at Coal Oil Point, off the coast of California – around 40,000 kg of methane per day.

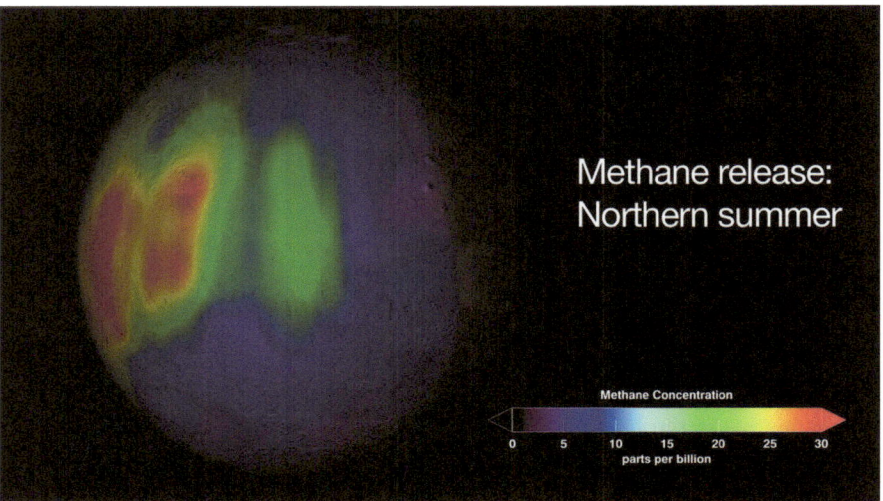

An image showing Martian methane distribution over one hemisphere in mid-summer, based upon spectroscopic observations made by NASA's Infrared Telescope Facility and the W.M. Keck telescope, both in Hawaii (Credit: NASA)

Measuring over 100 m in diameter, this squat dome shows all the hallmarks of being a Martian mud volcano. Now frozen, this feature may have produced the warm muddy conditions necessary for the proliferation of microbes and their release of methane into the Martian atmosphere (Credit: HiRISE, MRO, LPL (U. Arizona), NASA)

Astrobiologists are fascinated by the discovery of methane because of its implications for the existence of life. Much of Earth's methane (a so-called 'greenhouse gas') is released as a by-product of the digestion of nutrients by microorganisms called 'methanogens' present in rotting vegetation and in the stomachs of animals, notably cows and sheep. However, the gas can be produced as a consequence of non-biological activity; on Earth, for example, methane is released from deep inside the crust by mud volcanoes. Since the gas is rapidly broken down by ultraviolet light from the Sun, the short-lifetime of methane in Mars' atmosphere (just a few years at most) means that the gas is being replenished from active sources, whatever they may be. Should Martian microbes turn out to be the source, this methanogenic life is most likely to be found beneath the planet's surface where it is warm enough for the presence of liquid water and a carbon supply, both essential to life as we know it.

4.2 Temperature

Mars' average annual surface temperature is around −55°C. However, the planet's orbit is so eccentric that there is a variation in temperature throughout the Martian year of around 30°C at the subsolar point between the planet's perihelion and aphelion, producing a significant effect on the planet's climate; temperatures can plummet to as low as −140°C at the winter poles (50°C colder than the minimum recorded temperature on Earth) to around 27°C at the summertime subsolar point in areas with a low albedo (dark areas).

4.3 Atmospheric Pressure

On Mars' surface the atmospheric pressure varies from a near-vacuum of around 30 Pa on the highest point of Mars (the summit of Olympus Mons) where water only exists in the form of a solid or vapor, to more than 1,155 Pa at the deepest parts of the surface in the Hellas basin in the southern hemisphere. In comparison, Earth's average atmospheric pressure at sea level is 101.3 kPa – around 140 times that of Mars' average surface pressure. It is even possible for liquid water to temporarily exist in certain dark areas on the surface of Hellas Planitia, where the relatively high atmospheric pressure combined with occasional daytime temperatures exceeds the melting point of water ice. Indeed, much of Mars' northern hemisphere is consistently higher than the triple point of water, where the substance can exist either in gaseous, liquid or solid form.

4.4 Wind

Long before space probes imaged Mars up close and in detail, astronomers were sure that winds blew across the planet's surface. Astronomers observed changes in the brightness of the planet over time and suspected that wind-borne dust clouded the atmosphere. The presence of winds was overwhelmingly confirmed when Mariner 9 arrived at Mars in the middle of a huge dust storm. The Mariner and Viking spacecraft also revealed surface features that were obviously wind-formed (aeolian), including various types of dunes and windstreaks. Wind is the main process shaping the surface of Mars today.

4.5 Clouds

Clouds on Mars, though common, are usually much less pronounced than those visible in the skies of the Earth. Although Mars' atmosphere contains only a trace of water vapor – around one thousandth the amount of water vapor present in the Earth's atmosphere – this is sufficient to cause clouds to form. The atmospheric temperature and pressure is usually close to saturation and water vapor clouds are produced. From the Earth, cloud features are usually seen as transient bright features, but their structure has been closely examined by space probes and classified into the following types:

1. Lee waves, or orographic clouds, which form in the lee (the side opposite wind direction) of large obstacles like as ridges, craters, mountains and volcanoes, where wavelike oscillations are produced in the Martian atmosphere and producing clouds when moisture-laden air is forced upwards over these high points. Striking bright orographic clouds, brilliant and large enough to be seen through amateur telescopes, frequently form on the leeward side of Olympus Mons and the other Tharsis volcanoes.
2. Wave clouds, which appear as rows of linear clouds, most often found at the edges of the polar caps. These usually appear when the polar caps emerge into the Martian spring Sun, when the frozen carbon dioxide sublimes; winds are produced that sweep from the poles at high speed (up to 400 km/h).
3. Cloud streets which show a double periodicity, appearing as linear rows of bubbling, cumulus-like clouds.
4. Streaky clouds which have a direction without displaying periodicity.
5. Fog or ground hazes usually occur in low-lying areas such as valleys, canyons and craters, forming during dawn and dusk. Large patches of fog also can also form during the morning in low-lying areas, such as the canyons of Valles Marineris. Ground haze is occasionally produced by atmospheric dust.
6. Plumes, which are elongated clouds that appear to have a source of rising material; they are often composed of dust particles rather than water vapor. A remarkable example can be seen annually for a short time over one of the Tharsis volcanoes, Arsia Mons, as southern autumn ends and winter is about to begin; a spiraling cloud of fine dust, carried up the volcano's slopes by warm air, which reaches heights of up to 30 km.

A cloudy Mars. Image centred on Lunae Planum, Tharsis at left, Meridiani Planum at right, north polar cap visible at top (Credit: NASA/JPL)

Orographic clouds over the volcanic peaks of the central Tharsis region (Credit: NASA/JPL/MGS)

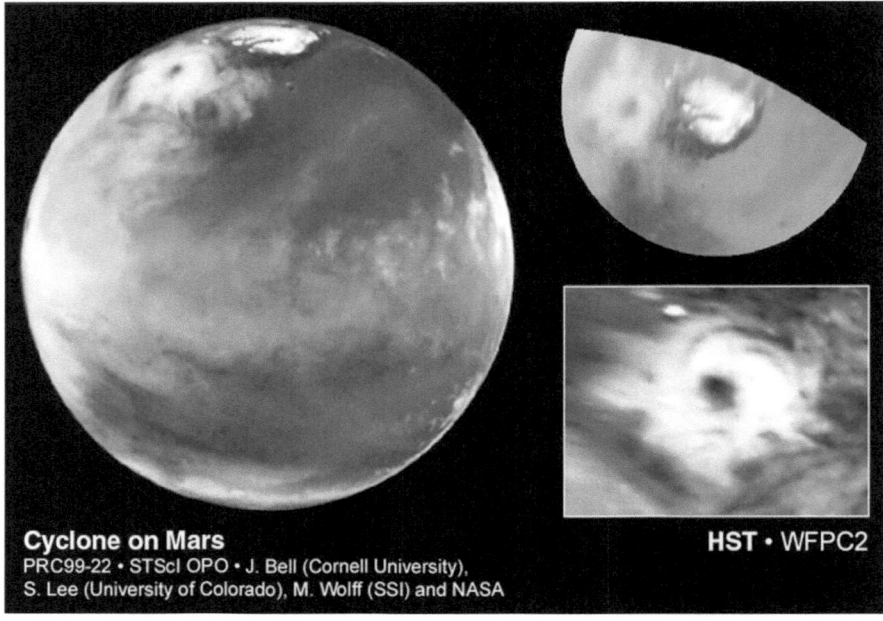

Cyclone on Mars
PRC99-22 • STScI OPO • J. Bell (Cornell University),
S. Lee (University of Colorado), M. Wolff (SSI) and NASA

HST • WFPC2

This image of a cyclonic storm near Mars' north polar ice cap was secured in April 1999 by the Hubble Space Telescope. The inset images at right show close-up views of the storm and surrounding areas. Measuring around 1,600 km across with a central eye of some 300 km, the cyclone abated within a few days (Credit: J. Bell (Cornell), S. Lee (Univ. Colorado), M. Wolff (SSI), et al., NASA)

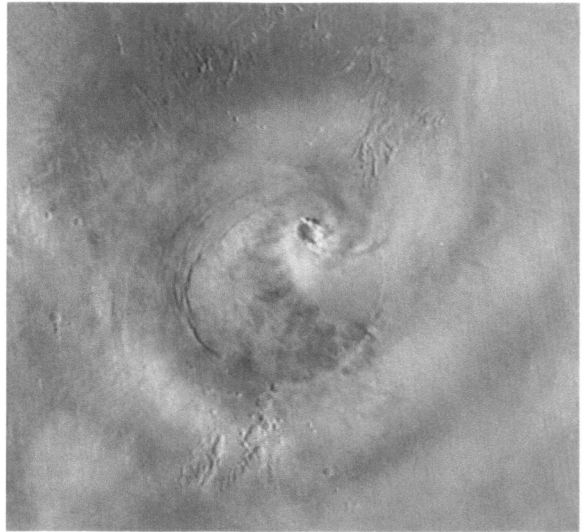

Spiral dust cloud over Arsia Mons, imaged during late southern autumn on Mars in April 2003 (Credit: NASA/JPL/Malin Space Science Systems)

Wispy clouds in the skies over Meridiani Planum (Credit: NASA/JPL)

Water frost also lightly coats the surface during the long chill of Martian winter. Mars' low atmospheric pressure prevents water from existing as a liquid on the surface; when the temperature rises above freezing point, water ice sublimates and turns directly into vapor. There is no doubt that long ago in the past Mars' atmospheric pressure was at times high enough to allow water to flow across its surface, forming shallow seas and lakes which were fed by rivers.

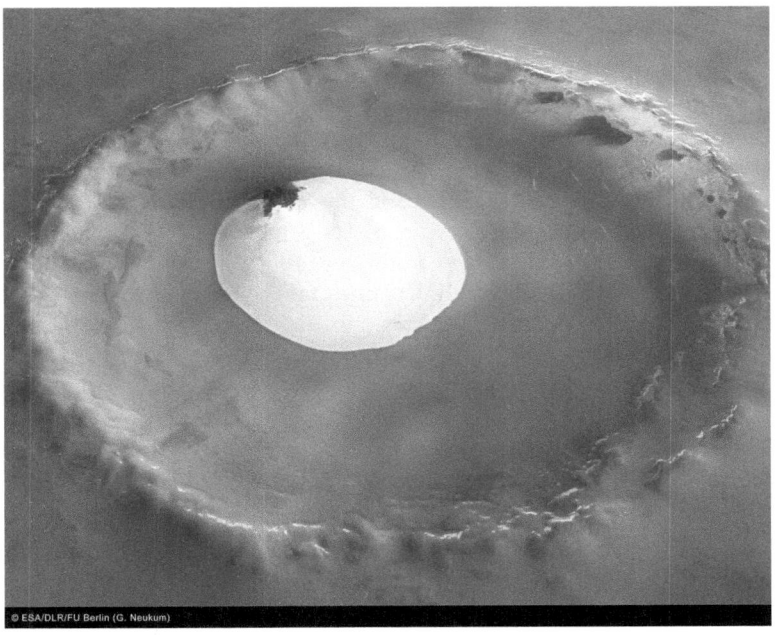

A 30 km crater in the Martian arctic region shows a pond of water ice on its floor and a frosted rim (Credit: ESA/Mars Express)

A thin layer of water ice frost (a few microns thick) coating the local surface, imaged from the plains of Utopia Planitia (Credit: Viking/NASA)

The dustiness of Mars' atmosphere, caused by wind whipping up the fine-grained surface material and holding it in suspension, produces a light brown or orange colored sky. In-situ space probe data indicates that these atmospherically-suspended dust particles are around 1.5 μm across. Tiny red colored dust particles (as fine as the particles making up cigarette smoke) suspended in Mars' atmosphere reflect sunlight and cause the sky to appear a light pinkish-brown color – the intensity of the color varies with the amount of dust in the atmosphere. On rare occasions when regional atmospheric conditions are perfectly calm, and the dust has settled out of the Martian skies, the color might turn into a deep blue, like the color of the sky seen from the summit of a terrestrial mountain on a clear day.

As Mars' atmosphere is heated by the Sun, the winds produced loft dust high into the air. Large amounts of dust are capable of causing Mars' features to appear muted, occasionally obscuring them from view altogether, on a regional or even global scale, for several weeks on end. The fine grains of dust storms reflect around 25% of sunlight falling on them, so they appear bright in comparison with the planet's darker desert features, which have an albedo of around 10%. Astronomers call these events dust storms, and although they may appear dramatic, on the planet's surface they are far less severe than a typical dust storm in one of Earth's deserts. Since the amount of solar energy received by Mars varies with season and distance from the Sun, there is a pattern to the Martian dust storms; they tend to be most severe when Mars approaches perihelion, when the planet receives 20% more solar energy than its yearly average.

Sunset on Mars, imaged from Gusev crater by the Spirit rover in May 2005. Note the dimming caused by dust in the Martian atmosphere (Credit: NASA)

Around a dozen major planet-wide dust storms have taken place since the perihelic apparition of 1877 – the famous apparition in which Giovanni Schiaparelli first observed 'canali' and Asaph Hall discovered the two tiny Martian satellites, Phobos and Deimos. On arrival at Mars in 1971, Mariner 9 was unable to view the planet's surface because it was obscured by a major global dust storm. The last one took place in 2001, when Mars was entirely covered by a hazy blanket which lasted for around 3 months.

In mid-2003, as Mars was approaching its perihelic opposition, several regional dust storms were observed, the largest of which spilled out from the Hellas basin and extended southwards across Syrtis Major to cover an area of around 600,000 km^2 within less than a week. Dust storms also brewed up at the end of 2003, threatening to converge and become a global phenomenon. The accumulation of dust poses a threat to landers and roving probes reliant on solar energy for their power.

26 June 2001   4 September 2001

From its vantage point in Earth orbit, the Hubble Space Telescope imaged Mars before (left) and during the great dust storm of 2001 which engulfed the planet for months (Credit: J. Bell (Cornell), M. Wolff (Space Science Inst.), Hubble Heritage Team (STScI/AURA), NASA)

Atmospherics, Meteors, and Magnetic Field

An orbital view by Mars Global Surveyor in 2003, showing a dust storm (at top) sweeping over Acidalia Planitia in the planet's northern hemisphere. The south polar cap is at the bottom of the image (Credit: Malin Space Science Systems, MGS, JPL, NASA)

In the summer of 2007 dust storms blocked much of the sunlight from reaching the solar panels of both the Spirit and Opportunity rovers, occasionally producing 99% obscuration of the Sun. This image from Opportunity shows a view of the increasing dimming of the skies above Victoria Crater (Credit: Mars Exploration Rover Mission, Cornell, JPL, NASA)

On a small scale, dust-devils frequently whip up and zip across the Martian plains. Dust devils are formed when air heated by a warm, poorly conducting desert surface, bubbles upward into colder air, dragging surrounding warm air upwards and causing a spinning vortex. The rotating column moves across the surface, picking up surface dust. Martian dust devils may rise to 8 km, but they only last several minutes as they strip off the terrain's brighter surface dust coating and leave behind dark winding trails. The phenomenon may look severe, but in comparison to terrestrial dust devils they are very mild.

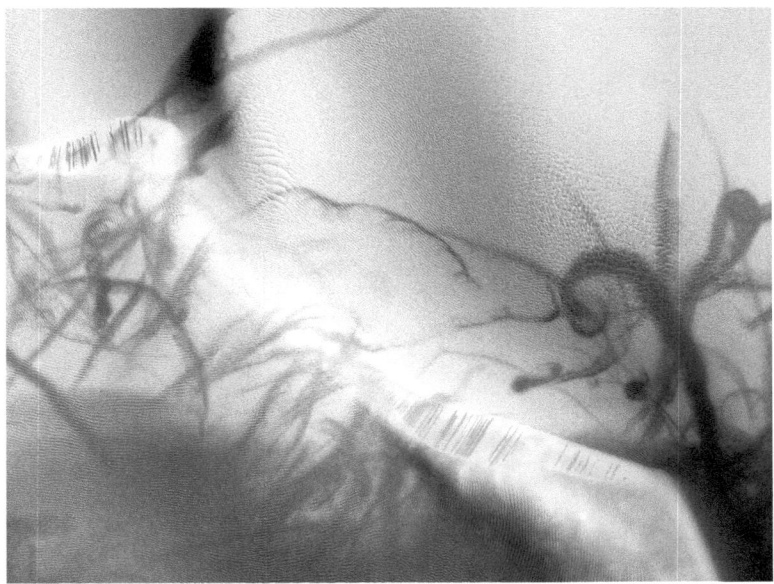

A high-resolution image of dust devil trails taken by the HiRISE camera on Mars Reconnaissance Orbiter (Credit: HiRISE, MRO, LPL (U. Arizona), NASA)

4.6 Martian Meteors

In 2005 the first image of a meteor streaking across Mars' sky was taken by the Spirit rover; an analysis of the image by one team of scientists even suggested that the meteor originated from comet 114P/Wiseman-Skiff.

Spirit rover's 2005 image of the first meteor detected in Mars' atmosphere. The object is travelling from left to right (Credit: Nature/NASA/Spirit/F. Selsis et al.)

Mars' orbit is approached by around four times as many comets as the Earth's (a good proportion of which are Jupiter Family Comets), so it is more than likely that the planet's orbit intersects a number of meteoroid streams deposited in their wakes. The Earth itself passes through dozens of meteoroid streams; as the sand grain-sized dust particles enter the atmosphere and burn up through friction at heights between 50 and 120 km, they produce annual meteor showers such as the Perseids, Lyrids, Geminids and Leonids which peak on certain dates. Owing to the larger mean scale height of the Martian atmosphere, meteoroids are thought to burn up over Mars at a similar altitude, producing meteors of similar magnitude to those observed from the Earth. We don't know the dates of Mars' annual meteor showers – that would require a dedicated panoramic night watch from a number of sites around the planet or continuous monitoring of the planet from orbit throughout at least a Martian year – but they doubtless exist. It is thought, for example, that Mars' orbit crosses that of the orbital stream of meteoroids produced by comets 1P/Halley, 13P/Olbers 79P/du Toit-Hartley.

As they burn up in the atmosphere, metallic particles within meteoroids are ionized, forming a short-lived trail of plasma, which is radio-reflective and detectable from the ground or from orbit. A team of scientists led by Armargh University produced a model that predicted the occurrence of six Martian meteor showers arising from comet 79P/du Toit-Hartley between 1997 (when Mars Global Surveyor began orbital observations) to 2005. After cross-referencing the predictions and the probe's observations of activity in Mars' ionosphere, data from April 2003 revealed an ionospheric disturbance took place at the same time and at the same height as the predicted meteor maximum. Future determinations of cometary orbits and their associated Mars-intersecting meteoroid streams, along with visual and radio observations from on and above Mars will doubtless uncover many more Martian meteor showers.

4.7 Meteorites on Mars

Following their superheated descent through the relatively thin Martian atmosphere, larger and more substantial meteoroids will survive to hit the planet's surface, producing a small crater on impact with the soil. A number of these meteorites have already been discovered here and there by the rovers Spirit and Opportunity; relatively fresh meteorites stand out from the rest of the surface features because they are un-weathered and have a distinct shape, color and (with nickel-iron meteorites) surface texture.

Between January 2005 and September 2010 the Mars rover *Opportunity* discovered six nickel-iron meteorites on the surface of Meridiani Planum, beginning with the first, 'Heat Shield Rock' (around 24 cm in diameter) and ending with 'Oileán Ruaidh' (about 45 cm); 'Block Island' (around 60 cm across and 30 cm high) is the largest meteorite to have been found. In addition, two nickel-iron meteorites were discovered by the Spirit rover, 'Allan Hills' (and 'Zhong Shan'), both in close proximity to each other, and a number of possible stony meteorites have been identified.

It must be noted that when referring to 'Martian meteorites' astronomers are usually talking about objects discovered on the Earth which are thought to have come from Mars (as described earlier in this chapter), a rare class of meteorite of which only 56 examples have been found.

'Heat Shield Rock' (officially named the Meridiani Planum meteorite), the first meteorite to have been discovered on the surface of Mars (Credit: NASA/JPL/Cornell)

4.8　Martian Meteorites

More than 53,000 well-documented meteorites have been found on Earth. Most of these space rocks originate from the Asteroid Belt, but there exists a very special sub-class – those originating from Mars. Only 99 such meteorites have been identified, of which all but one are stony (achondritic) and fall into the SNC Group, named after the shergottites, nakhlites and chassignites (2). There's little doubt that the SNC meteorites come from Mars – their elemental and isotopic compositions are very similar to in-situ analysis by space probes of Martian rocks and atmospheric gases. Martian meteorites are thought to have been ejected from Mars by asteroidal or cometary impact and were fortuitously captured by the Earth some time later.

Shergottites, 83 of which are known, are igneous mafic (rich in iron and magnesium) in origin and appear to have crystallized as recently as 180 million years ago – a remarkably (some would maintain impossibly) recent date, prompting much debate and active research, although the shergottite age paradox is currently unresolved.

Known nakhalites are 13 in number. The first to be identified, a 10 kg specimen, fell in El-Nakhla, Alexandria, Egypt in 1911 and is said to have killed a dog. Igneous and rich in augite, they formed from basaltic magma – perhaps in the large volcanic regions of Tharsis, Elysium or Syrtis Major Planum – some 1.3 billion years ago, and liquid water is thought to have suffused them around 620 million years ago. The impact that ejected the Nakhlites from Mars took place less than 11 million years ago, and their arrival on Earth took place within the last 10,000 years.

Two chassignites are known – one, the Chassigny meteorite, fell at Chassigny in France in 1815, while the other, NWA 2737, was found in Morocco in 2000; the two have strong similarities with each other. Composed mostly of olivine, cassignites formed in the Martian mantle at around the same time as the nakhalites.

In a class of its own is meteorite ALH 84001, found in Antarctica's Allan Hills in 1984. Thought to have crystallized from molten rock more than four billion years ago, ALH 84001 was shocked by impacts while on Mars and blasted off the surface by an impact about 15 million years ago; it landed on Earth around 13,000 years ago. The material making up the meteorite may therefore have originated during the early Noachian Period at a time when liquid water existed on Mars, as waterborne carbonate minerals occupy cracks in its structure. In 1996 evidence for fossilized life within ALH 84001 was claimed to have been detected in the form of tiny bio-deposited magnetite and tiny tubular structures that were speculated to be nanobacterial fossils. Although the origin of such features in ALH 84001 has not been proven, many scientists remain optimistic that simple life may have developed in Mars' past, and may indeed have survived to this day.

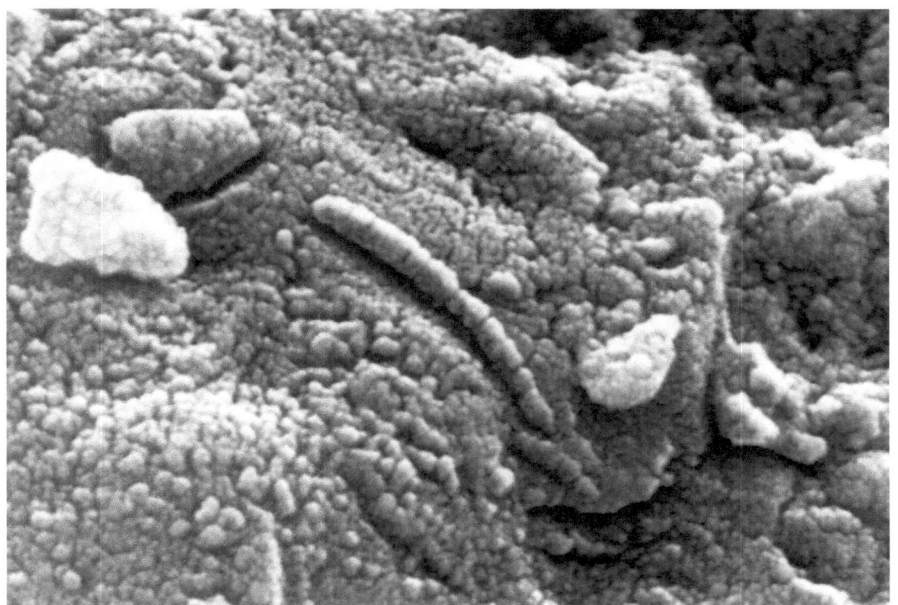

Electron microscope image of Martian meteorite ALH 84001 showing enigmatic structures which some think represent microfossils (Credit: NASA)

4.9 Magnetic Field

We are all used to the effects of the Earth's dipolar magnetic field – compasses point towards the magnetic pole, and colorful aurorae are produced when charged particles streaming from the Sun spiral down towards the polar regions along magnetic field lines and collide with molecules in our atmosphere. Like a protective shield, the magnetosphere deflects the solar wind, keeping us safe from its damaging effects. There would be no use on Mars for a magnetic compass, and auroral displays never grace the Martian skies because the planet has no appreciable global magnetic field – it's certainly less than 1,000 times less than the strength of the Earth's.

Earth's magnetic field is produced by dynamo effects around 3,000 km beneath the surface in a continuously moving layer of fluid molten iron in the planet's outer core; Mars lacks a molten core and its ability to create a magnetic field through dynamo effects ceased just a few hundred million years after the planet's formation. However, Mars' original magnetic field was perhaps around one-tenth the current strength of that of the Earth, and this magnetized parts of the crust up to the time that the planet's core solidified.

Evidence of magnetized rocks and portions of the crust indicates that Mars lost its global magnetic field a few hundred million years after its formation, at around the same time that the Solar System's late heavy bombardment period was in full swing. The shutdown of the Martian magnetic field four billion years ago may have been due to the planet's internal cooling and the consequent solidification of its core, gradually putting the brakes on the dynamo effect. It has been suggested that the reasons for such an early cessation of a planetary magnetic field was more dramatic and was a consequence of around half a dozen particularly huge asteroidal impacts during the late heavy bombardment period. Such impacts generated tremendous amounts of heat, and as the crust and mantle cooled heat flow in the planet's core was disrupted and the core rapidly lost its heat and ability to flow.

Impact craters formed after this time show no signs of magnetism – they obliterated portions of the pre-existing magnetized crust and the molten rock they produced didn't assume any traces of magnetism. In 1997 Mars Global Surveyor detected weak magnetism in the most ancient parts of the planet's crust – these were the remnants of the planet's original magnetic field. It has been shown that the field changed polarity several times during the first few hundred million years after Mars was formed.

Chapter 5

The Martian Moons

Phobos and Deimos, the two satellites of Mars, are potato-shaped city-sized lumps
of rock orbiting in near-circular paths almost directly above the Martian equator.
Both moons are made of very dark material and display cratering. Because of their
low surface brightness and composition it was once widely considered that both
Phobos and Deimos were captured asteroids, but models show that the likelihood
of captured asteroids assuming equatorial, near-circular orbits is extremely low. It
is possible that they accreted from orbiting debris, perhaps from material blasted
from Mars itself by large impacts.

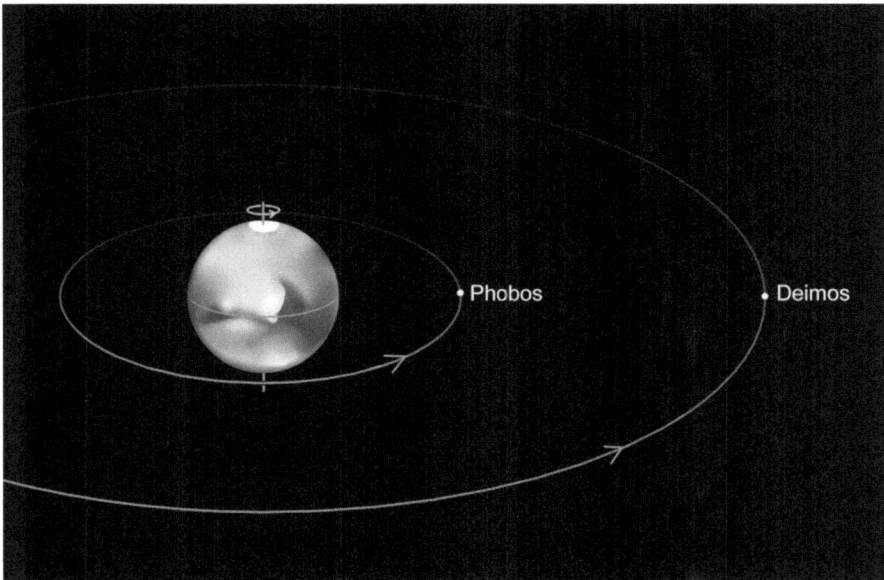

The orbits of Phobos and Deimos, shown to scale (Credit: Grego)

P. Grego, *Mars and How to Observe It*, Astronomers' Observing Guides,
DOI 10.1007/978-1-4614-2302-7_5, © Springer Science+Business Media New York 2012

Phobos (left) and Deimos, shown to scale (Credit: NASA/JPL)

Phobos, the larger of the pair, orbits at a distance of just 9,377 km from the center of Mars (5,981 km above the surface) in a period of just 7 h 39 m – the closest planetary satellite and the only one to orbit its primary faster than the primary revolves. Seen from the equator on Mars' surface it rises in the west and sets in the east 4 h 20 m later, only to rise again after just 11 h. On rising, Phobos is around 8.5 arcminutes across, but its closeness to Mars means that as it transits the meridian directly overhead it has grown to 12 arcminutes in diameter, more than one-third the apparent diameter of our own Moon as seen from the Earth. At progressively higher northern or southern latitudes Phobos crosses the sky in an even shorter period as the arc it describes moves nearer the horizon, until at latitudes greater than 70°N or S it can't be seen at all. Its apparent brightness as seen from Mars' surface varies with its distance from the observer, its apparent phase and the distance between Mars and the Sun; at Martian perihelion it can appear as bright as magnitude −9.5.

Phobos has an irregular shape, measuring 26.8 × 22.4 × 18.4 km, with an average diameter of 22.2 km and a surface area of 6,100 km² (about the same area as Ireland's County Galway). Its albedo of 0.071 is very low, making it one of the Solar System's least reflective objects. Spectroscopically resembling C- or D-type asteroids, Phobos' composition is similar to carbonaceous chondrite meteorites, but its density is too low for it to be made of solid rock. It must therefore be made of quite porous material, with possible icy constituents lying deep beneath a fine-grained regolith which may be more than 100 m thick. How Phobos, with its negligible gravity, has retained such a regolith, remains something of a mystery.

Phobos is dominated by Stickney, a deep crater measuring 9 km across (around one third of Phobos' length). Much of Phobos' surface is striated with parallel grooves and interconnected crater chains with diameters ranging from a few tens to several hundreds of meters. The grooves and crater chains are centered on the orbital leading end of Phobos (near the crater Stickney) and fade towards the trailing end.

Once speculated to have originated as a result of the Stickney impact, it is now thought that these odd looking features were formed by a multitude of impacts from material thrown out by impacts on Mars. More than a dozen families these features have been identified, each linked with a Martian impact event.

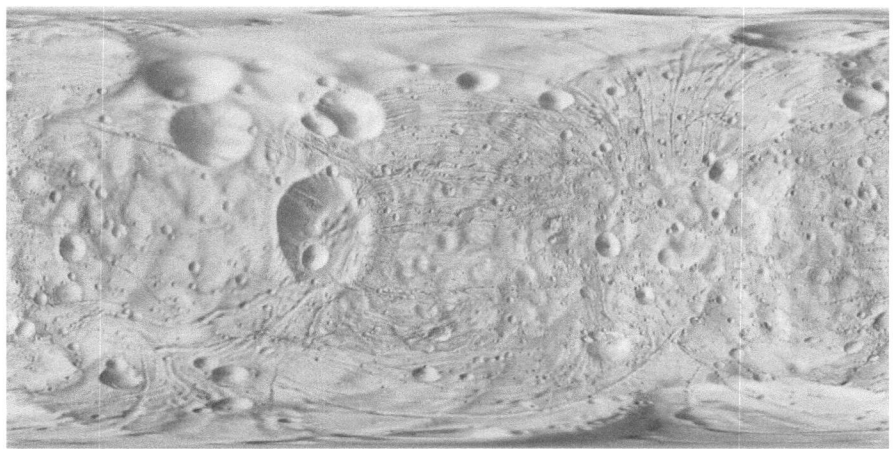

Modified Mercator map of Phobos (Credit: NASA/USGS)

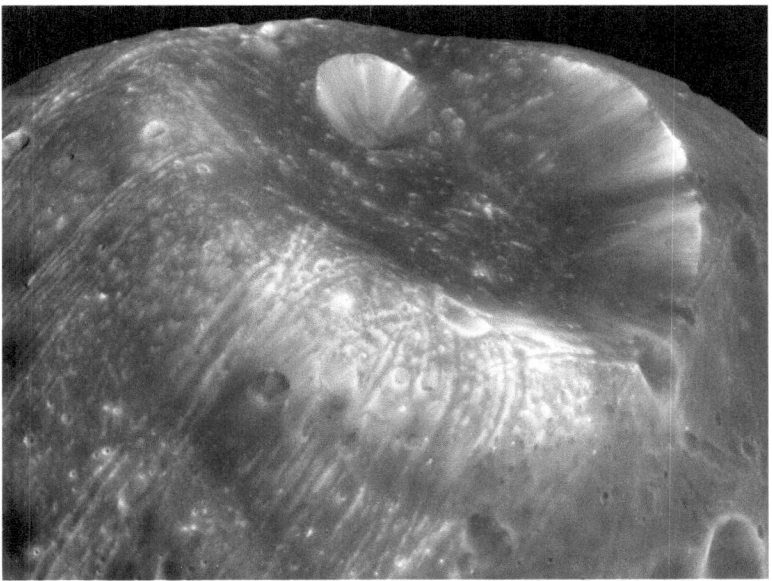

Close-up of Stickney, the largest crater on Phobos (Credit: NASA/USGS)

Phobos is doomed to destruction in the very near future, astronomically speaking. Its orbit is so low that tidal forces are reducing its orbital distance, making it spiral downwards. In less than 8 million years Phobos will orbit at a height of just 7,100 km, passing beyond what is known as the Roche Limit, below which it will lose its own gravitational integrity. The little moon will consequently break up to form a ring system around Mars.

Orbiting some 23,460 km from the center of Mars (2 km above the planet's surface) in a period of 30 h 18 m, Deimos is in a much 'safer' place than Phobos – tidal forces are actually working to increase the moon's distance from Mars. Viewed from the Martian equator Deimos rises slowly in the east and sets in the west some 60 h later, during which time it passes near (or transits) the Sun and is 'buzzed' by Phobos on a number of occasions. With an apparent diameter of just 2.5 arcminutes when it soars overhead, it would be difficult for a keen-sighted person to distinguish Deimos' phase without optical aid. From latitudes greater than 82.7° N and S, Deimos never rises above the Martian horizon.

Deimos is an irregularly shaped object measuring $15 \times 12.2 \times 10.4$ km, with a surface area of 1,400 km^2 (about the same area as the Greek island of Rhodes). Like Phobos, Deimos' spectrum resembles a C- or D-type asteroid. It is less obviously cratered than Phobos – most of its craters are smaller than 2.5 km across – and its surface is blanketed with a thick regolith of soil and rock (perhaps 100 m deep) which fills in most of the craters and makes the moon appear a very smooth object.

Modified Mercator map of Deimos (Credit: NASA/USGS)

Chapter 6

A Topographic Survey of Mars

The following survey of the entire surface of Mars is divided into four regions of equal area which cover both northern and southern hemispheres in 90° wide longitudinal sections, progressing west from the Martian prime meridian. These regions are the Acidalia-Marineris-Argyre Region (0–90°), the Arcadia-Tharsis-Sirenum Region (90–180°W), the Utopia-Elysium-Cimmeria Region (180–270°W) and the Vastitas-Sabaea-Hellas Region (270–360°W). The polar areas are discussed in more detail above. The survey is largely topographically descriptive, but additional information has been included in terms of setting some features against a geological perspective.

P. Grego, *Mars and How to Observe It*, Astronomers' Observing Guides,
DOI 10.1007/978-1-4614-2302-7_6, © Springer Science+Business Media New York 2012

Region One (centre 45°W)　　Region Two (centre 135°W)

Region Three (centre 225°W)　　Region Four (centre 315°W)

Viking-based hemispheric albedo maps showing the four regions of Mars (north at top) (Credit: NASA/Google Earth/Grego)

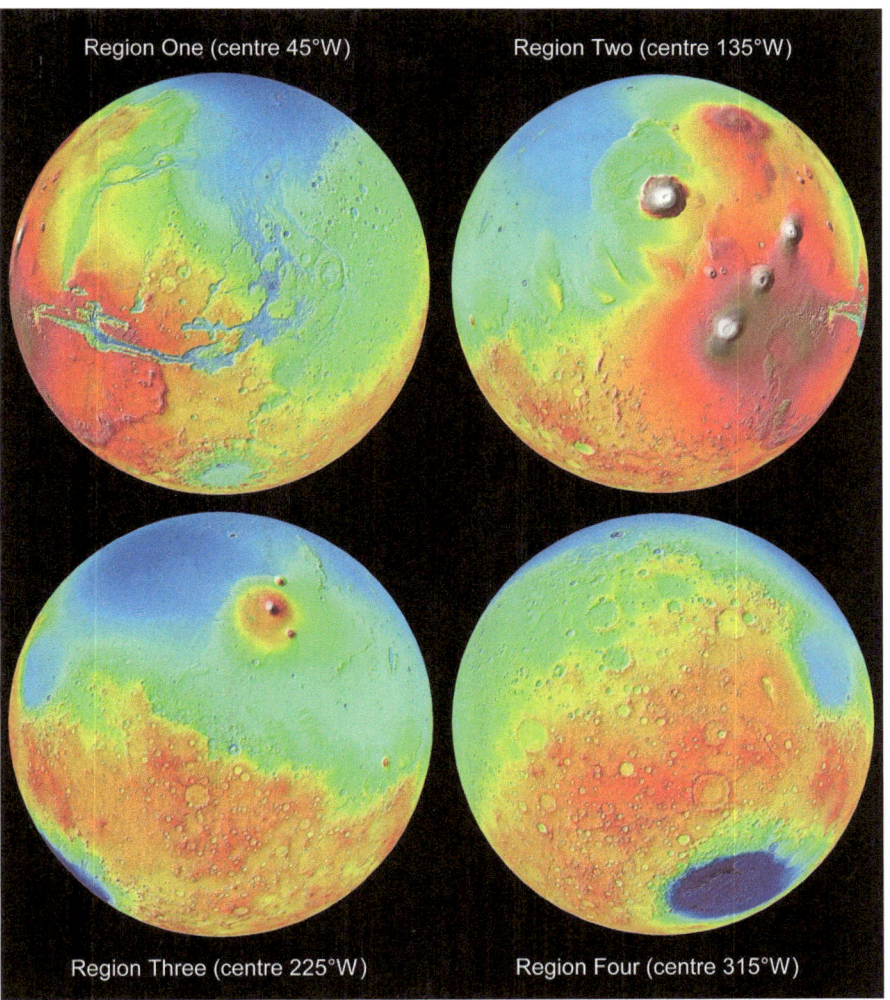

Region One (centre 45°W)   Region Two (centre 135°W)

Region Three (centre 225°W)   Region Four (centre 315°W)

Color-coded topographic hemispheric maps showing the four regions of Mars (north at top) (Credit: NASA/Google Earth)

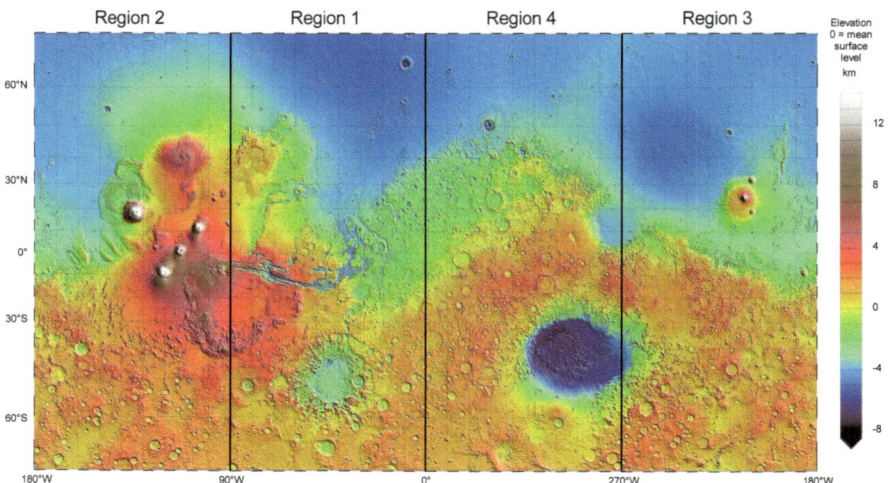

Color-coded topographic map of Mars, showing the four regions described in the following text (Credit: NASA/Grego)

Each of these four Martian regions is surveyed in a very general north to south, westward trend, using the larger topographic features as the main points of reference in the text. Features inside or close to the main reference features are surveyed in a rough anticlockwise trend from north. In order to help the flow of the narrative, these rules are of a general nature, and there are a number of diversions and some overlaps with adjoining regions where necessary. Initial references to each particular feature are in bold face and are followed by the latitude and longitude (in parentheses) of the feature's central point, to the nearest degree; this information is often accompanied by the feature's main dimensions, its diameter/length and/or height. The text is not exhaustive, but it covers all the main features on Mars and mentions some interesting ones besides. Additional topographic features are also mentioned in the text above, but to minimize repetition many of these are not covered in this survey.

Feature names, named feature coordinates and the dimensions of named features have been derived from the US Geological Survey's Astrogeology Research Program website Gazetteer of Planetary Nomenclature (URL in appendix).

Accompanying each regional map is a labeled map showing the features mentioned in this survey of Mars' topographic features, along with a lettered and numbered key to the features described. Providing the text is read with occasional reference to these maps and the images that accompany it, the reader will be in little danger of losing themselves on this fascinating world.

6.1   A Note on Co-ordinates

Each map and image in this survey is presented with north at top, west at left. To avoid confusion, the maps cover the same areas that are discussed in the guide to telescopically observable albedo features in Part Two of this book. Note, however, that the nomenclature is different between the two surveys. In the following topographical survey the official IAU planetary nomenclature is used, while the observer's survey in Part Two of this book uses Antoniadi's classic nomenclature which is based on albedo features and commonly used to this day by visual observers and imagers. It will be found that much of the IAU nomenclature relating to large features pays respect to the pre-Space Age nomenclature; to give just one example, the location of Solis Planum corresponds with the dusky area of Solis Lacus in the old maps.

A westward longitude, increasing from 0° to 360°, has been adopted for all co-ordinates in this book; the same system, sharing the same prime meridian, is used on albedo maps, for visual work and in most 'official' maps prior to 2000. Long used by visual observers, this positive westward longitude – a so-called 'planetographic' system using surface-mapped co-ordinates – was used for charting features observed by the Viking probes of the 1970s.

However, the US Geological Survey and NASA later adopted a 'planetocentric' system – using co-ordinates measured from the planet's center – with longitude increasing eastward; this was to be used in future mapping and imaging by space probes. Both planetographic and planetocentric systems were approved by the IAU in 2000, but maps produced after this date increasingly use the planetocentric system.

Another system, based on the one we use on Earth, gives both positive west and east bearings – 0 to 180°W (west from the prime merdian to the antemeridian) and 0–180°E (east from the prime meridian to the antemeridian). It can be a trifle confusing, but it is simple to convert eastern co-ordinates to western ones – just subtract the east figure from 360°.

6.2   Feature Types

Catena (catenae): A chain of craters of roughly similar size. 16 named.

Cavus (cavi): Rimless, irregular-shaped deep hollow. 16 named.

Chaos (chaoses): An area of jumbled or hummocky 'chaotic' terrain. 26 named.

Chasma (chasmata): Large canyon, steep sided valley or trough. 25 named.

Collis (colles): A small hill or group of small hills. 17 named.

Crater (craters): Generally circular depression, most often found with a raised rim, ejecta deposits and secondary impact features (impact crater). 990 named.

Dorsum (dorsa): A ridge. 33 named.

Fluctus (fluctūs): A flow-like feature. 2 named.

Fossa (fossae): A narrow, linear trough. Groups are often parallel or cross-cut. 56 named.

Labes (labēs): A landslide feature. 5 named.

Labyrinthus (labyrinthi): An area of interconnected valleys or canyons. 6 named.

Lingula (lingulae): Lobate or tongue-like extension to a plateau. 5 named.

Mensa (mensae): A flat-topped elevated feature with steep sides. 28 named.

Mons (montes): A mountain. 45 named.

Palus (paludes): Classical name for a smooth area of medium albedo (adopted on modern IAU maps). 4 named.

Patera (paterae): A complex or irregularly-shaped volcanic crater with low relief and radial channels. 15 named.

Planitia (planitiae): A plain with a lower elevation than its surroundings. 10 named.

Planum (plana): A smooth plateau with steep sides. 31 named.

Rupes (rupēs): A cliff or escarpment. 23 named.

Scopulus (scopuli): A lobate cliff or escarpment. 13 named.

Sulcus (sulci): A ditch-like feature, always found in near-parallel groups. 13 named.

Terra (terrae): A large elevated land mass. 11 named.

Tholus (tholi): A rounded hill or small mountain. 18 named.

Unda (undae): A wave-like dune field. 5 named.

Vallis (valles): Most often a sinuous valley created by fluvial processes, or in some cases a larger valley formed by tectonic activity. 136 named.

Vastitas (vastitates): A large lowland plain. 1 named.

6.3 Region 1: Acidalia-Marineris-Argyre Region (0–90°W)

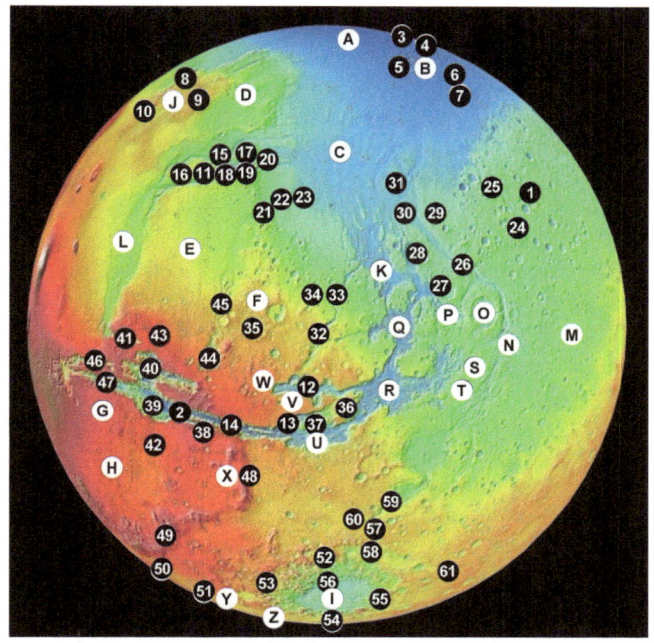

Map of region 1, centred on the equator at 45°W, showing features mentioned in the descriptive text. Key (in order of first mention in the survey): *A*, Vastitas Borealis; *B*, Acidalia Planitia; *C*, Chryse Planitia; *D*, Tempe Terra; *E*, Lunae Planum, Sacra Dorsa; *F*, Xanthe Terra; *1*, Becquerel; *2*, Valles Marineris; *G*, Sinai Planum; *H*, Solis Planum, Solis Dorsa; *I*, Argyre Planitia; *3*, Lomonosov; *4*, Kunowsky; *5*, Acidalia Colles; *6*, Elath, Lota; *7*, Arandas; *8*, Mareotis Fossae; *9*, Tempe Fossae; *J*, Ascuris Planum; *10*, Reykholt, Gonnus Mons, Tanaica Montes, Tanais Fossae, Baphyras Catena; *11*, Kasei Valles; *K*, Chryse Chaos; *12* Ganges Chasma; *13*, Capri Chasma; *14*, Coprates Chasma; *L*, Echus Fossae, Echus Chasma; *15*, Lobo Vallis; *16*, Sacra Mensa; *17*, Sharonov; *18*, Lunae Mensa; *19*, Nilokeras Fossae; *20*, Rongxar; *21*, Xanthe Montes, Vedra Valles; *22*, Maumee Valles, Xanthe Scopulus; *23*, Iamuna Dorsa; *24*, Trouvelot; *25*, Mawrth Vallis; *M*, Meridiani Planum; *26*, Ares Vallis; *N*, Iani Chaos; *O*, Aram Chaos; *27*, Galilaei, Barsukov, Silinka Vallis; *P*, Hydaspis Chaos; *28*, Sagan, Masursky; *29*, Oxia Colles; *30*, Oraibi; *31*, Wahoo, Yuty, Wabash; *Q*, Hydraotes Chaos; *32*, Orson Welles; *33*, Shalbatana Vallis; *34*, Nanedi Valles, Hypanis Valles; *35*, Mutch; *R*, Aurorae Chaos; *S*, Margaritifer Terra; *T*, Margaritifer Chaos; *U*, Eos Chaos; *36*, Eos Mensa; *37*, Capri Mensa; *V*, Aurorae Planum; *W*, Ophir Planum, Ophir Catenae; *X*, Thaumasia Planum, Felis Dorsa; *38*, Coprates Catena; *39*, Melas Chasma; *40*, Candor Chasma, Ophir Chasma; *41*, Hebes Chasma; *42*, Melas Dorsa; *43*, Juventae Dorsa; *44*, Juventae Chasma; *45*, Maja Valles; *46*, Tithonium Chasma, Tithoniae Catenae; *47*, Ius Chasma, Louros Valles; *48*, Nectaris Fossae; *49*, Coracis Fossae; *50*, Thaumasia Fossae; *51*, Lowell; *52*, Nereidum Montes; *53*, Bosporos Rupes; *54*, Charitum Montes; *55*, Galle; *56*, Hooke; *57*, Bond, Uzboi Vallis; *58*, Hale; *59*, Holden; *60*, Nirgal Vallis; *Y*, Aonio Planum; *Z*, Argentea Planum; *61*, Lohse. Note that letters refer to features of extended area, ie., vastitas, terrae, planitiae, chaoses, paludes and plana, while numbers refer to all other features (Credit: NASA/Google Earth/Grego)

Region 1 is perhaps the most featuresome and variety-packed of the four regions covered in this survey. In the north, a large V-shaped wedge of relatively smooth lowlands, consisting of part of **Vastitas Borealis** (89.8°N, 0.0°W; 3,500 km), **Acidalia Planitia** (45.5°N, 25.5°W; 3,363 km) and **Chryse Planitia** (28.4°N, 40.3°W; 1,542 km) intrudes between two very different kinds of loftier terrain – namely, **Tempe Terra** (39.0°N, 70.6°W; 1,955 km), **Lunae Planum** (10.9°N, 65.5°W; 1,818 km) and **Xanthe Terra** (1.6°N, 48.1°W; 1,868 km) to the west, and the cratered uplands in the vicinity of **Becquerel** (22.3°N, 8.0°W; 171 km) to the east. Southern Chryse Planitia is linked by a series of valleys to the chasmata at the far eastern end of the great **Valles Marineris** (14.2°S, 58.6°W; 3,761 km long) system. South of Valles Marineris lies the upland regions of **Sinai Planum** (13.7°S, 87.8°W; 901 km) and **Solis Planum** (26.7°S, 89.7°W; 1,811 km). **Argyre Planitia** (50.2°S, 43.3°W; 893 km) dominates the central southern region.

Above 60°N, between 0° and 80°W, the lowlands of Vasitas Borealis deepen to their lowest levels, in some places 4 km below datum and lower on the floors of some of the larger impact craters – in **Lomonosov** (65.3°N, 9.2°W; 131 km) for example, which lies just south of the Martian Arctic Circle, the largest crater to be found in this particular region of Mars. Four hundred and eighty kilometers due south of Lomonosov lies **Kunowsky** (56.8°N, 9.7°W; 67 km); together, the craters form a prominent duo in an otherwise bland landscape. Southwest of Kunowsky, the northern part of Acidalia Planitia displays the hilly terrain of **Acidalia Colles** (50.1°N, 23.1°W; 356 km). The area, thought to be a repository for considerable amounts of sub-surface ice, is known for its craters with lobate ejecta systems. Among these is **Elath** (45.9°N, 13.7°W; 13 km), **Lota** (46.4°N, 11.9°W; 15 km) and **Arandas** (42.7°N, 15.0°W; 25 km) with its remarkable 'mudsplash' pattern.

Tempe Terra, another region thought to contain considerable quantities of sub-surface ice, forms an extended upland apron to the northeast of the Tharsis highlands. Tempe Terra is distinctly higher in the west than in the east, and it is riven with the southwest-northeast trending rifts of the **Mareotis Fossae** (43.7°N, 75.3°W; 1,860 km long) and **Tempe Fossae** (39.9°N, 71.4°W; 2,000 km), some of the latter of which curve around the eastern border of **Ascuris Planum** (40.4°N, 80.8°W; 500 km). Steep cliffs, mensae and plana are to be found along the northern edge of Ascuris Planum (the western half of Tempe Terra). Standing proud here, a short distance west of the modified infilled crater **Reykholt** (40.5°N, 86.3°W; 53.2 km) is a group of mountains, including the solitary triangular massif of **Gonnus Mons** (41.3°N, 91.0°W; 57 km, 2,890 m above datum) and the elongated spine of **Tanaica Montes** (39.5°N, 91.1°W; 177 km, 1,980 m). South of Reykholt amid the **Tanais Fossae** (38.9°N, 86.6.°W, 166 km) is the elongated collapse valley **Baphyras Catena** (39.2°N, 84.2°W; 96 km).

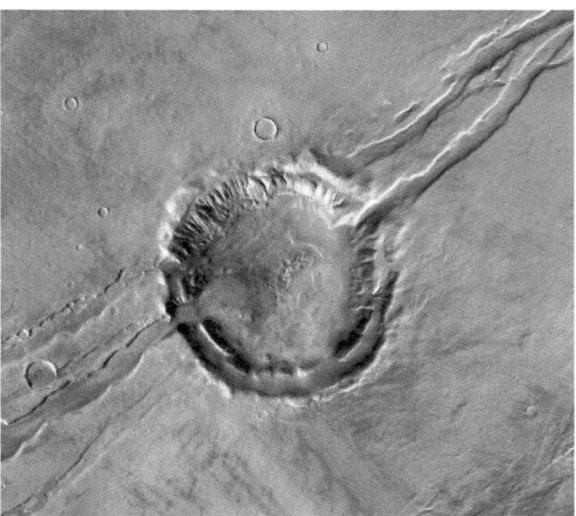

This unnamed crater (41.8°N, 77.2°W; 29 km) is cut through by one of the rifts of Tempe Fossae. Later, sediment has been carried along the floor of the fossa and deposited on the crater's floor (Credit: NASA/JPL/Grego)

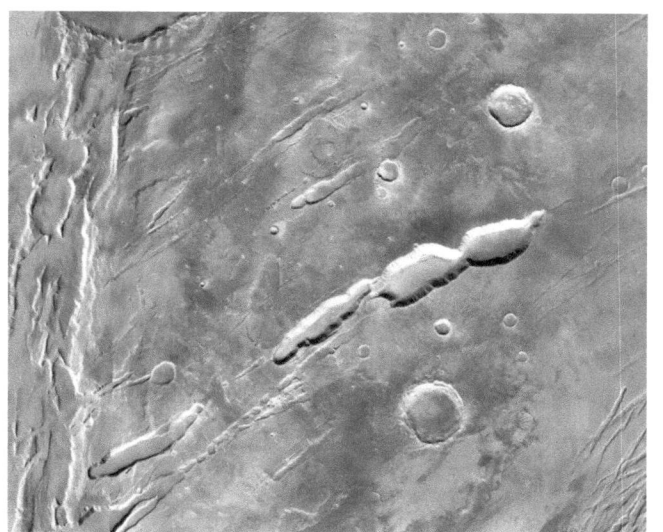

Baphyras Catena (centre) and the Tanais Fossae (left) (Credit: NASA/JPL/Grego)

Together, the adjoining wrinkled uplands of Lunae Planum, ridged by the **Sacra Dorsa** (9.8°N, 66.3°W; 1,630 km) and the cratered terrain of Xanthe Terra form a continent-sized 'island' whose perimeter is marked by the vast outflow channels of **Kasei Valles** (24.4°N, 65.0°W; 1,780 km) to the north, Chryse Planitia and **Chryse Chaos** (15.7°N, 35.7°W; 1,720 km) to the east, the rift valleys **Ganges Chasma** (7.9°S, 48.1°W; 584 km), **Capri Chasma** (9.8°S, 43.3°W; 1,275 km) and **Coprates Chasma** (13.3°S, 361.4°W; 966 km) to the south and the well-defined **Echus** lowlands to the west.

Kasei Valles, which once drained meltwaters from the west into Chryse Planitia, forms the largest system of outflow channels on Mars; its major northern component, **Lobo Vallis** (26.9°N, 61.2°W, 102 km) snakes north of the streamlined lenticular plateaux of **Sacra Mensa** (24.5°N, 68.1°W; 580 km) and **Sharonov** (27.0°N, 58.6°W, 102 km), while its southern branch proceeds south of these features, around **Lunae Mensa** (24.0°N, 62.6°W; 117 km) and proceeds north of **Nilokeras Fossae** (24.8°N, 57.8°W, 265 km) and out to the teardrop-shaped 'island' at the head of which lies the crater **Rongxar** (26.3°N, 55.5°W; 22 km).

Xanthe Montes (18.4°N, 54.5°W; 500 km, 1,620 m, below datum) form a mountainous arc around the southwestern edge of Chryse Planitia and represent part of one of the large rings which once surrounded the Chryse impact basin; the mountains are cut through by a number of outflow channels, including **Vedra Valles** (19.2°N, 55.6°W; 115 km) which extend onto the Chryse plains in the form of the **Maumee Valles** (19.5°N, 53.2°W; 350 km). In this area are visible more remnants of the Chryse multi-ring impact basin, including the **Xanthe Scopulus** (19.3°N, 52.6°W; 57 km) and **Iamuna Dorsa** (20.9°N, 50.4°W; 49 km).

The cratered uplands to the south of Acidalia Planitia contain several large craters, including Becquerel and **Trouvelot** (16.2°N, 13.1°W; 155 km), both of which have sizeable craters on their floors. A winding outflow channel, **Mawrth Vallis** (22.4°N, 16.5°W; 636 km) runs northwest from the flanks of Trouvelot. The slightly more elevated and less-cratered upland area further south is named **Meridiani**

Planum as it is located on the equator at 0° longitude, the Martian prime meridian. Several larger outflow channels are to be found to the west, including **Ares Vallis** (10.3°N, 25.8°W; 1,700 km) and **Tiu Valles** (15.7°N, 35.7°W; 1,720 km).

Ares Vallis winds its way northwest from the rough broken landscape of **Iani Chaos** (2.8°S, 17.5°W; 434 km), around the eastern perimeter of **Aram Chaos** (2.6°N, 21.5°W; 277 km) and broadens out to Chryse Planitia, and is joined along its route by numerous smaller side valleys. One such small tributary originates within Aram Chaos – really, a large crater with a rough floor – and cuts through Aram's eastern wall. Other craters in this area also display wall-cutting valleys, including one in the southwest wall of **Galilaei** (5.6°N, 27.0°W; 137 km) which joins **Hydaspis Chaos** (3.2°N, 27.1°W; 355 km) and another, known as **Silinka Vallis** (8.9°N, 29.2°W; 140 km) in the northern wall of **Barsukov** (7.9°N, 29.1°W; 72 km). North of Aram and the mid-way point of Ares Vallis there is an unnamed east-facing lobate scarp some 340 km long which cuts through a number of pre-existing craters – this may be a remnant of one of the Chryse impact basin's rings. On the opposite (west) side of Ares Vallis, the crater **Sagan** (10.7°N, 30.7°W; 98 km), with its clean-cut appearance, central peak ring and hummocky ejecta makes a nice comparison with **Masursky** (12.0°N, 32.4°W; 118 km) to its northwest; Masursky is an older crater whose southern wall has been completely breached, its floor covered with material and later eroded into a mass of small angular mesas, while a valley, part of the Tiu Valles, cuts through its north wall.

Aram Chaos (circular feature near centre) with Ares Vallis to its east, Iani Chaos to its south and Hydaspis Chaos to its west (Credit: NASA/JPL/Grego)

As it broadens out in the north Ares Vallis flows by numerous streamlined 'islands' and craters to the west of a low, hummocky plain occupied by the **Oxia Colles** (21.5°N, 26.7°W; 569 km); among these are **Oraibi** (17.2°N, 32.4°W; 33 km) at the mouth of Ares Vallis, and further north the trio of **Wahoo** (23.3°N, 33.6°W; 67 km), **Yuty** (22.2°N, 34.2°W; 20 km) and **Wabash** (21.4°N, 33.7°W; 42 km), the latter three displaying lobate ejecta blankets.

Joining Ares Vallis in southern Chryse are the broad outflows of Tiu Valles to the south; these link with the jumbled lowlands of Chryse Chaos further south, which in turn is connected to the remarkable patchwork landscape of **Hydraotes Chaos** (0.8°N, 37.4°W; 418 km). The cratered plains of Xanthe Terra to the west host a number of meandering valleys. Spilling out from the northeastern rim of **Orson Welles, Shalbatana Vallis,** makes its way to Chyrse Chaos; running roughly parallel to this, the intricate narrower sinuous valleys of **Nanedi Valles** (4.9°N, 49.0°W; 508 km) link with similar systems in **Hypanis Valles** (9.5°N, 46.7°W; 231 km). One of the largest impact craters in this region of Mars, **Mutch** (0.6°N, 55.3°W; 211 km) lies some 500 km west of Orson Welles.

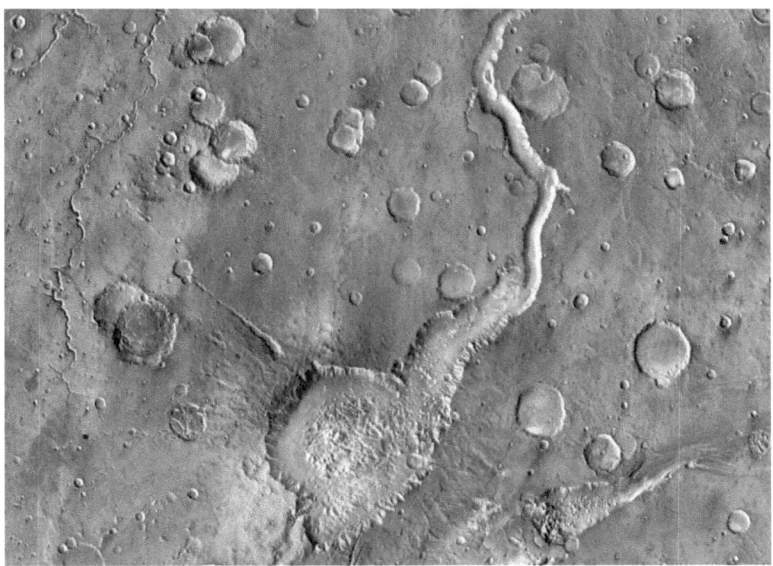

Shalbatana Vallis spills out from the crater Orson Welles. The much narrower Nanedi Valles can be seen at the far left. Topographic image (Credit: NASA/JPL/Grego)

Moving on to the grander sub-equatorial valley systems in and around Valles Marineris, we begin with **Aurorae Chaos** (8.9°S, 35.3°W; 750 km), a sunken lowland region to the west of **Margaritifer Terra** (4.9°S, 25.0°W; 2,600 km) whose own chaotic region, **Margaritifer Chaos** (8.6°S, 21.6°W; 390 km) is at a higher level than Aurorae Chaos, with a surface displaying a less advanced stage of subsidence. Aurorae Chaos covers a wide area and its surface is more colles-like than many other chaotic regions; it is linked with valley systems in the north and west.

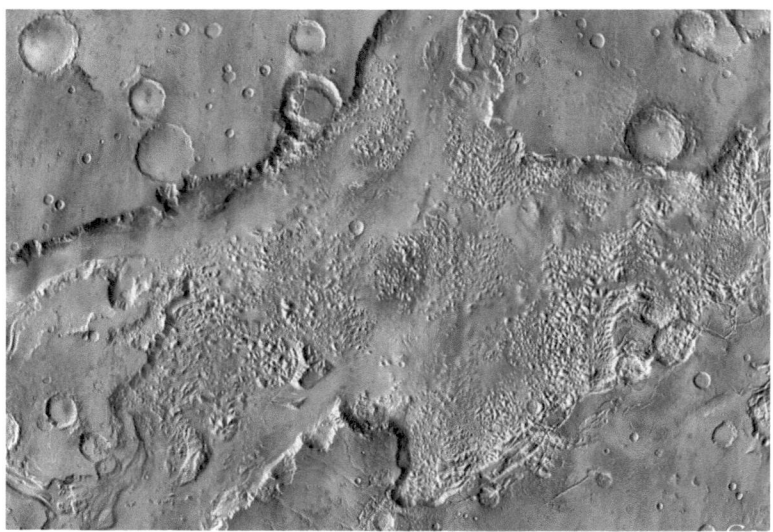

Aurorae Chaos. Topographic image (Credit: NASA/JPL/Grego)

Aurorae Chaos branches in the west; the northern branch extends past Capri Chasma and into Ganges Chasma, while a broader southern branch extends into Eos Chasma and **Eos Chaos** (16.6°S, 48.9°W; 490 km). The two branches outline the large 'island' plateux of **Eos Mensa** (10.9°S, 42.2°W; 390 km) and **Capri Mensa** (13.9°S, 47.4°W; 275 km) and then rejoin, the floor deepening below datum and below the surrounding highlands, with **Aurorae Planum** (10.4°S, 49.2°W; 600 km) to the north; this marks the entrance of Coprates Chasma – spanning 16° of longitude in an almost west–east orientation it is the longest rift valley of the Valles Marineris system. Hundred kilometers wide and 9,000 m deep in places, Coprates Chasma has a well-preserved floor which hasn't been infilled with great piles of various layered deposits as have many other Martian valleys.

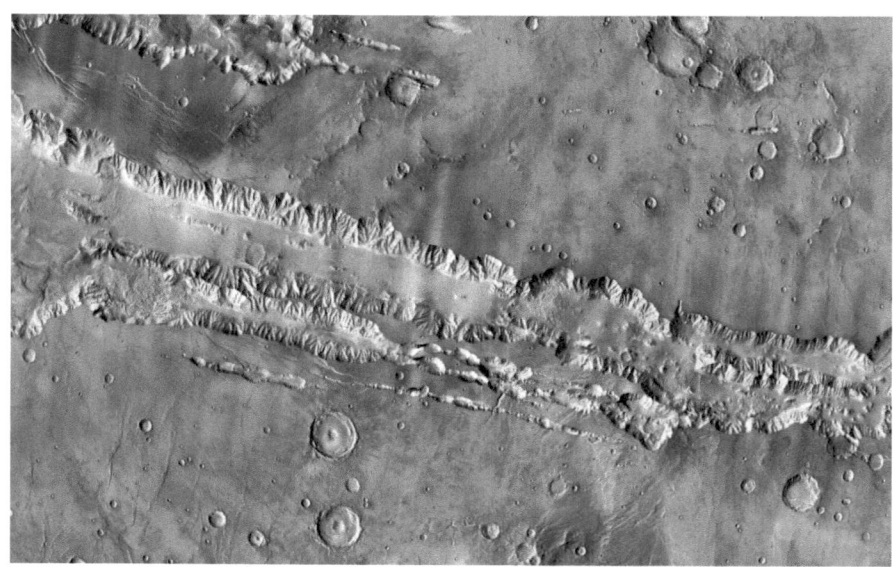

Coprates Chasma, nearly 1,000 km long. Topographic image (Credit: NASA/JPL/Grego)

North of Coprates Chasma lies **Ophir Planum** (8.7°S, 57.5°W; 650 km), a relatively smooth upland plain streaked with a number of west–east chain craters, the **Ophir Catenae** (9.5°S, 59.4°W; 577 km). South of Coprates Chasma bulges the ridged highland of **Thaumasia Planum** (24.5°S, 64.3°W; 650 km) with **Coprates Catena** on its northern border; here, and in the neighbouring sprawling plains of eastern Solis Planum, there are the extensive north–south trending ridges of **Felis Dorsa** (21.9°S, 65.9°W; 783 km), **Solis Dorsa** (23.1°S, 79.8°W; 878 km) and **Melas Dorsa** (18.3°S, 71.7°W; 560 km); similar ridge systems are also present north of Ophir Chasma in the form of **Juventae Dorsa** (0.1°S, 71.4°W; 519 km) these ridges all forming at around the same time by forces of crustal compression.

Around half way along the Valles Marineris, Coprates Chasma broadens into **Melas Chasma** (10.3°S, 72.7°W; 547 km); this extends north into the sunken landscapes of **Candor Chasma** (6.5°S, 68.9°W; 813 km) and **Ophir Chasma** (4.0°S, 72.5°W; 317 km). North lies the detached **Hebes Chasma** (1.1°S, 76.2°W; 319 km) and, further east, **Juventae Chasma** (3.5°S, 61.4°W; 320 km) with its drainage channels flowing north across eastern Lunae Planum into **Maja Valles** (12.5°N, 58.3°W; 1,516 km).

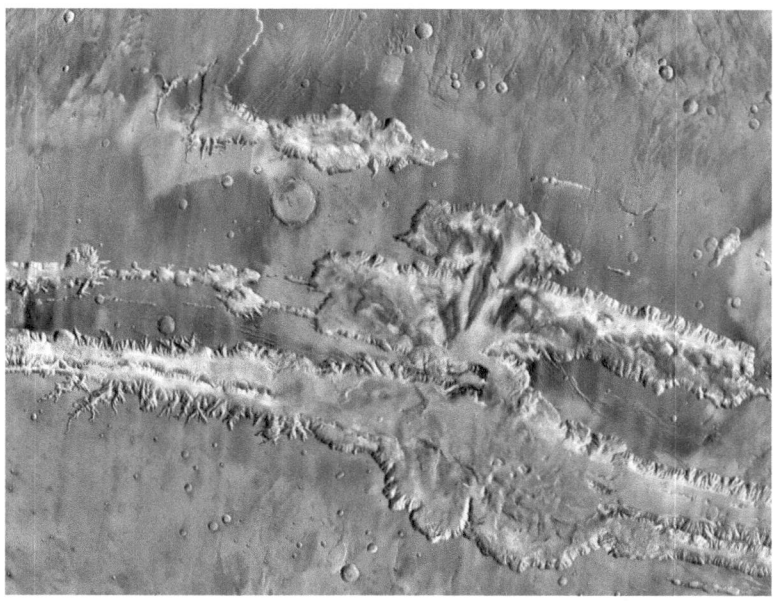

Western valles marineris, with Melas Chasma, Candor Chasma, Ophir Chasma and Hebes Chasma; the valley narrows further west along Ius Chasma and Tithonium Chasma. Topographic image (Credit: NASA/JPL/Grego)

West of Melas Chasma the valley system narrows, where parallel valleys become **Tithonium Chasma** (4.6°S, 84.7°W; 810 km) in the north and **Ius Chasma** (6.9°S, 85.8°W; 938 km) in the south, separated by a narrow plateau upon which is the impressive collapse pit crater chain of **Tithoniae Catenae** (5.3°S, 82.4°W; 567 km). The steep walls of Ius Chasma show many smaller side valleys, notable among them the **Louros Valles** (8.4°S, 82.0°W; 517 km) on its south side.

Marking the opposite (south) border of Solis Planum, a very large unnamed highland ridge, 3,500 km long and 500 km wide in places, is cut through by north–south trending rupes and fossae in the south; the ridge runs around the southern and eastern border of Thaumasia Planum, where the fossae are east–west trending. Features along this ridge include (from east to west) **Nectaris Fossae** (24.3°S, 57.4°W; 650 km), **Coracis Fossae** (35.6°S, 70.6°W; 780 km) and northern extensions of **Thaumasia Fossae** (47.2°S, 92.7°W; 1,028 km). Nestled at the southeastern edge of Thaumasia Fossae, the large double-ringed crater **Lowell** (52.0°S, 81.4°W; 203 km) makes an impressive sight.

The 'bullseye' crater lowell. Topographic image (Credit: NASA/JPL/Grego)

Argyre Planitia is one of Mars' largest and best-preserved multi-ring impact basins. Several rings are visible surrounding a relatively smooth inner plain some 700 km across; between this and the outer ring, marked in the north by **Nereidum Montes** (38.6°S, 44.0°W; 1,130 km), **Bosporos Rupes** (42.9°S, 57.6°W; 507 km) in the northwest and **Charitum Montes** (58.0°S, 40.2°W; 850 km) in the south, there is a jumbled mass of hills and mountains with radial trends, a feature originally imparted by impact but later exacerbated by fluvial erosion. Two sizeable craters overlap the inner ring, namely double-ringed **Galle** (50.9°S, 30.9°W; 230 km) in the east and **Hooke** (44.9°S, 44.4°W; 139 km) in the north. Via crater **Bond** (32.9°S, 36.0°W; 111 km) the **Uzboi Vallis** (29.5°S, 37.1°W; 366 km) outflow channel links **Hale** (35.8°S, 36.5°W; 149 km) on the northern outer ring of the Argyre basin with crater **Holden** (26.1°S, 34.0°W; 154 km). Feeding into this valley from the cratered plain north of the Argyre basin is the narrow meandering 'river bed' **Nirgal Vallis** (28.1°S, 42.0°W; 496 km).

The cliffs of Bosporos Rupes are found on the cratered eastern plains of Bosporos Planum; two a lower, considerably less cratered plains lie to its south, **Aonia Planum** (57.7°S, 79.0°W; 650 km) and **Argentea Planum** (69.8°S, 68.0°W; 1,750 km). The uplands south and east of Argyre Planitia, up to the prime meridian, show little of more than ordinary interest, other than impact craters of various sizes; a chance alignment of four large impact craters runs from Galle to **Lohse** (43.4°S, 16.8°W; 156 km) over a distance of 875 km.

Argyre Planitia dominates the landscape in the southern central section of region 1. Topographic image (Credit: NASA/JPL/Grego)

6.4 Region 2: Arcadia-Tharsis-Sirenum Region (90–180°W)

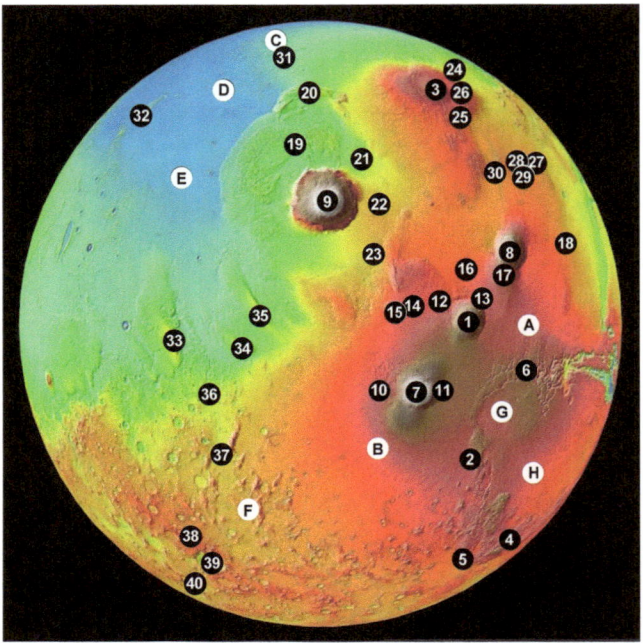

Map of Region 2, centred on the equator at 135°W, showing features mentioned in the descriptive text. Key (in order of first mention in the survey): A, Tharsis; 1, Tharsis Montes, Pavonis Mons; B, Daedalia Planum; C, Vastitas Borealis, Milankovic; D, Arcadia Planitia; E, Amazonis Planitia; F, Sirenum Terra; 2, Claritas Rupes, Claritas Fossae; G, Syria Planum; H, Solis Planum; 3, Alba Mons, Alba Patera; 4, Coracis Fossae; 5, Thaumasia Fossae; 6, Noctis Labyrinthus, Noctis Fossae; 7, Arsia Mons; 8, Ascraeus Mons; 9, Olympus Mons, Pangboche, Karzok; 10, Aganippe Fossa, Arsia Sulci; 11, Oti Fossae; 12, Pavonis Sulci; 13, Pavonis Fossae, Pavonis Chasma; 14, Ulysses Tholus, Ulysses Patera; 15, Biblis Tholus, Biblis Patera; 16, Poynting; 17, Ascraeus Chasmata; 18, Tharsis Tholus; 19, Lycus Sulci; 20, Acheron Fossae; 21, Cyane Sulci; 22, Sulci Gordii; 23, Gigas Sulci; 24, Tantalus Fossae; 25, Ceraunius Fossae; 26, Phlegethon Catena, Acheron Catena; 27, Uranius Mons; 28, Uranius Tholus; 29, Ceraunius Tholus; 30, Tractus Fossae, Tractus Catena; 31, Arcadia Dorsa; 32, Erebus Montes; 33, Eumenides Dorsum; 34, Amazonis Mensa; 35, Gordii Dorsum; 36, Mangala Valles; 37, Mangala Fossa; 38, Sirenum Fossae; 39, Newton; 40, Copernicus. Note that letters refer to features of extended area, ie., terrae, planitiae, chaoses, paludes and plana, while numbers refer to all other features (Credit: NASA/Google Earth/Grego)

The predominant feature of this region is the gigantic crustal uplift of **Tharsis**, a name given to the general area centered at 0.0°N, 100.0°W (an albedo feature). For our purposes here, Tharsis will refer to the entire uplifted region to which Tharsis is central, including the Alba region to the north and the **Daedalia Planum** (21.8°S, 128.0°W; 1,800 km) and **Solis Planum** regions in the south. Around 4,400 km across, Tharsis straddles the equator between around 60° and 140°W and covers an area of more than 15 million square kilometers; it rises to an average of 10 km above datum and in places exceeds 20 km high.

North of Tharsis there are the Vastitas Borealis lowlands, while to its west sprawl the lowland plains of **Arcadia Planitia** (46.7°N, 168.0°W; 2,200 km) and **Amazonis Planitia** (24.8°N, 164.0°W; 2,800 km). Tharsis' southwestern slopes are crossed by numerous ridges parallel to the Tharsis bulge and various fossae separating the cratered highlands of **Sirenum Terra** from the smoother sloping heights of

Daedalia Planum, while the southern part of Tharsis is bisected by a vast arcuate mountainous ridge separating Daedalia Planum, Solis Planum and the southern cratered uplands. In the west this ridge is cut into by the enormous west-facing escarpment of **Claritas Rupes** (25.7°S, 105.4°W; 924 km) and the north–south trending **Claritas Fossae** (31.2°S, 104.1°W; 2,050 km), while its southern heights are streaked by the **Coracis Fossae** (35.6°S, 80.6°W; 780 km) and northern extensions of **Thaumasia Fossae** (47.2°S, 92.7°W; 1,028 km). In the north, the ridge is met by the western extensions of Valles Marineris, the jumbled network of valleys of **Noctis Labyrinthus** (6.9°S, 102.2°W; 1,263 km).

Tharsis is Mars' most extensive volcanic province, home to some of the Solar System's mightiest shield volcanoes. With a relatively young Amazonian Period surface, Tharsis is free of large impact craters; what large craters it has are volcanic calderas. Lying in a southwest-northeast line, the three volcanoes making up the main **Tharsis Montes** (1.2°N, 112.5°W; 1,840 km) are **Arsia Mons** (8.3°S, 120.1°W; 435 km across, 17.9 km above datum), **Pavonis Mons** (0.8°N, 113.4°W; 375 km, 14.0 km) and **Ascraeus Mons** (11.8°N, 104.5°W; 460 km, 18.2 km).

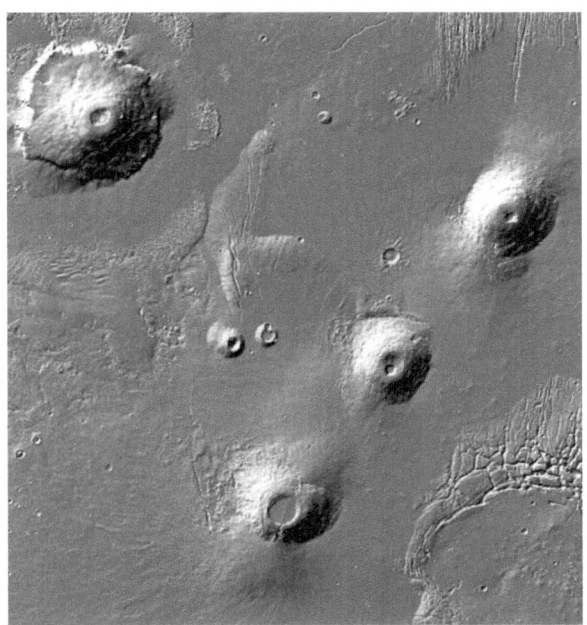

Tharsis Montes, with Olympus Mons at top left, Noctis Labyrinthus at bottom right. Topographic image (Credit: NASA/JPL/Grego)

Arsia Mons, the most southerly of the Tharsis Montes, has a footprint of 148,000 km^2 and a volume 30 times that of Hawaii's Mauna Loa, the Earth's largest volcano. Rising above the surrounding highlands, it is topped by a circular caldera 110 km wide. There is some evidence that sub-surface glaciers exist on Arsia Mons; a series of ridges and lobes overlying older surface features west of the caldera has been interpreted as moraines left by a receding glacier. A number of collapse features can be found on the flanks of the volcano, including catenae extending south from the caldera rim and channels on its northern slopes; several pits and cave entrances (averaging 100 m across) have also been identified. A large north–south trench, **Aganippe Fossa** (8.2°S, 126.2°W; 532 km) cuts through the furrowed terrain of **Arsia Sulci** (6.0°S, 128.9°W; 475 km) to the west of Arsia Mons, while the parallel grooves of **Oti Fossae** (9.2°S, 116.8°W; 370 km) can be found to the east.

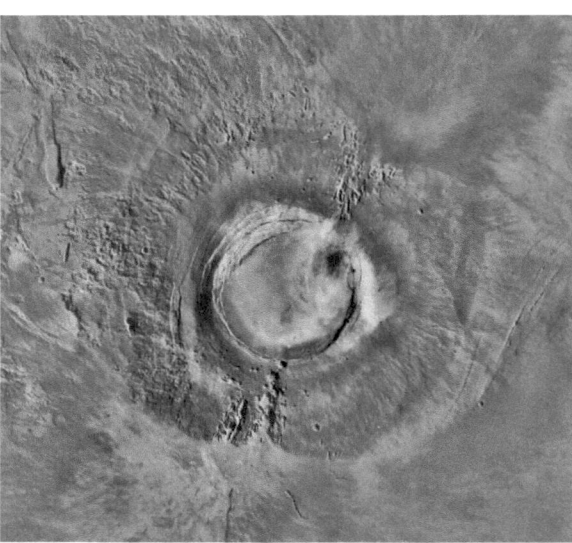

Arsia Mons, the most southerly of the Tharsis Montes. Topographic image (Credit: NASA/JPL/Grego)

Central to the Tharsis Montes, lying squarely on the Martian equator, is Pavonis Mons. This shield volcano is topped by an old caldera 80 km wide, inside the southwest of which lies a smaller, younger and more clearly defined circular crater 47 km in diameter. Like its neighboring shields, its outer flanks are cut through by sulci, fossae and chasmata, namely **Pavonis Sulci** (4.0°N, 116.5°W; 429 km) to the west, while to the north lie the parallel ditches of **Pavonis Fossae** (4.1°N, 111.5°W; 168 km) and **Pavonis Chasma** (2.7°N, 111.2°W; 45 km).

Pavonis Mons, the central volcano of the Tharsis Montes. Topographic image (Credit: NASA/JPL/Grego)

A number of interesting features lie in the general vicinity of Pavonis Mons. Five hundred kilometers to its west can be found two small volcanoes, **Ulysses Tholus** (2.9°N, 161.6°W; 102 km) and **Biblis Tholus** (2.7°N, 164.6°W; 172 km) each of which is capped by a 50 km caldera, **Ulysses Patera** and **Biblis Patera**. An unnamed carter 30 km across – one of the few large impact craters in this entire region – lies across the northern flanks of Ulysses Tholus. Another, larger impact crater, **Poynting** (8.3°N, 112.9°W; 74 km) is to be found 380 km north of Pavonis Mons.

Poynting, one of the largest impact craters in the Tharsis region. Topographic image (Credit: NASA/JPL/Grego)

Ascraeus Mons, the tallest and most northerly of the large Tharsis Montes, has a complex caldera consisting of a young deep central crater 30 km across overlying four smaller pre-existing calderas which gives the system a multi-lobed flower-like appearance. It is thought that the central caldera is about 100 million years old, while the others formed about 200, 400, and 800 million years ago. In common with the other Tharsis Montes, Ascraeus Mons is surrounded by a series of arcuate rounded terraces formed as a result of crustal compression and thrust-faulting. Also in common with its sister volcanoes, Ascraeus Mons has a fan-shaped area of rough ground bounded by arcuate ridges lies on its western flanks; this is thought to be a made up of glacial deposits. At the northeastern and south-western base of the volcano are groups of sinuous valleys with large heads, including **Ascraeus Chasmata** (8.7°N, 155.7°W; 110 km); in some cases these outflow channels cut across narrow arcuate valleys, concentric to the volcano, caused by crustal tension.

Ascraeus Mons, the most northerly of the Tharsis Montes. Topographic image (Credit: NASA/JPL/Grego)

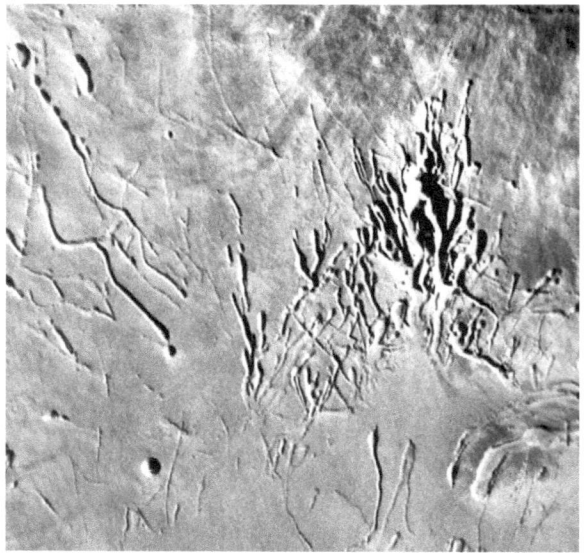

Ascraeus Chasmata, southwest of Ascraeus Mons. Topographic image (Credit: NASA/JPL/Grego)

Seven hundred and seventy five kilometers east of Ascraeus Mons, the small volcano **Tharsis Tholus** (13.4°N, 90.8°W; 158 km) rises 8 km above the surrounding plains. It consists of a central caldera, 45 km wide, which is superimposed upon two converging arcuate ridges which appear to be the remnants of older calderas.

Just northwest of the main Tharsis bulge lies the mighty **Olympus Mons** (18.4°N, 134.0°W; 648 km) the largest shield volcano in the Solar System. With a base covering an area of around 300,000 km² (about the same as Italy), Olympus Mons rises to 21 km above Mars' mean surface level (22 km above datum). Averaging 6°, the gradient of its outer slopes slightly increases towards the summit so that in profile it has a slightly convex appearance. A flat summit, offset from the center of the mountain by around 20 km to the east, is topped by a conjoined cluster of six calderas measuring 85 km from northeast to southwest and 60 km from northwest to southeast. It is thought that the calderas date from between 350 and 150 million years ago, all forming within 100 million years of each other. The largest caldera formed as a single lava lake, fed by a magma chamber estimated to lie around 32 km beneath the caldera's floor.

Although it looks impressive in images taken from orbit, Olympus Mons wouldn't look nearly as awesome from the ground; an astronaut standing on its slopes would find it difficult to perceive the volcano's enormity – indeed, viewed from its base, the summit of Olympus Mons would lie far beyond an almost flat horizon.

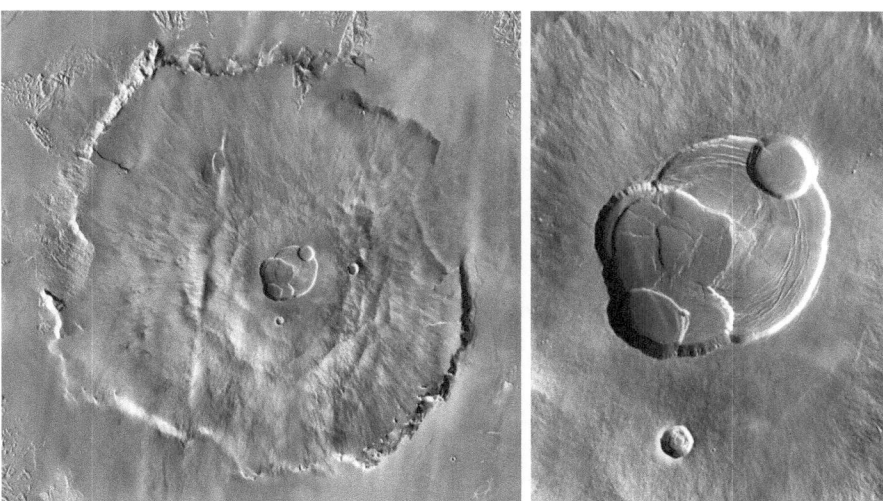

Olympus Mons and a close-up of its complicated caldera. Topographic image (Credit: NASA/JPL/Grego)

Two sizeable impact craters lie on Olympus Mons' upper slopes – **Pangboche** (17.0°N, 133.6°W; 10 km) in the south and **Karzok** (18.2°N, 131.9°W, 16 km).

In addition to being asymmetrical in shape, there are dissimilarities in the structures seen on either side of Olympus Mons. On the shallower northwestern slopes are found features formed by crustal tension, including slumping and normal faults producing several prominent escarpments, while compressional features such as dorsa and arcuate rounded terraces formed by thrust faults are to be found on the steeper southeastern flanks. Olympus Mons' base is defined by an enormous escarpment, in places 8 km high. Immediately at the base of this cliff there is a 'moat' which is some 2,000 m deep in the northwest; this feature, unique to Olympus Mons, is likely to have been caused by the sheer weight of the volcano pushing down on the crust.

Several extensive lobes of wrinkled and furrowed terrain, known as the Olympus Mons aureole, surrounds Olympus Mons; these vast aprons are thought to have formed through landslides on the steep outer border of the volcano. To the north and west, where the exposed aureole is at its most extensive (up to 700 km broad) are the **Lycus Sulci** (24.4°N, 141.1°W; 1,350 km across); this area of the Olympus Mons aureole nudges up to an 850 km long highland arc which is streaked by the **Acheron Fossae** (37.3°N, 137.8°W; 718 km). Northeast of Olympus Mons lies the crescent-shaped plateau of **Cyane Sulci** (25.5°N, 128.7°W, 340 km) and **Sulci Gordii** (18.7°N, 125.5°W; 400 km) to the east, while to the southeast, 340 km from the edge of Olympus Mons, lies **Gigas Sulci** (9.9°N, 127.8°W; 398 km).

Alba Mons (40.5°N, 109.6°W; 530 km), an enormous but squat shield volcano with very low slopes of around 3°, lies some 1,700 km northeast of Olympus Mons. Taking the area occupied by the volcano's outer lava flows as the area covered by the feature, it is actually around 3,000 km wide and 2,000 km from north to south, with an area of some 1,800,000 km^2 (equivalent to the area of Libya).

This volcano may have had a more exotic origin than most others; exactly antipodal to the Hellas impact basin, it has been suggested that seismic waves generated by the Hellas impact travelled around Mars and were focused in one specific spot, weakening the crust and allowing the volcanism that formed Alba Mons to commence.

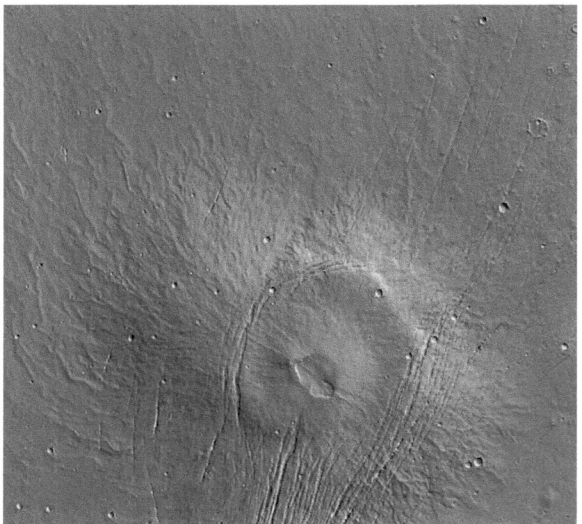

Alba Mons and its surrounding fossae. Topographic image (Credit: NASA/JPL/Grego)

At the summit of Alba Mons, the main caldera **Alba Patera** (39.80°N, 109.8°W; 136 km) contains a smaller, deformed caldera (50 km across) on its southern floor. At its deepest, the caldera's floor is 1,200 m below its rim. Alba Patera is more clearly defined in the west, where its steep scalloped wall rises to 500 m. A small, squat volcanic dome, 50 km across, complete with an unusual double-ringed feature like ripples in a pond, can be found on Alba Patera's eastern floor. Further out, streaming around the base of the 'official' Alba Mons, are numerous parallel fossae and catenae; flowing around the north, **Tantalus Fossae** (50.6°N, 97.5°W; 2,400 km), with **Ceraunius Fossae** (29.2°N, 109.0°W; 1,137 km) to the south, while between the two, west of Alba Mons, run **Phlegethon Catena** (38.9°N, 103.3°W; 391 km) and **Acheron Catena** (37.3°N, 101.0°W; 491 km).

Five hundred kilometers east of Ceraunius Fossae, on the Tharsis plains beyond Alba Patera, there's a trio of small volcanoes – **Uranius Mons** (26.8°N, 92.2°W; 62 km wide, 4,853 m high), **Uranius Tholus** (26.1°N, 97.7°W; 62 km, 4,290 m) and **Ceraunius Tholus** (24.0°N, 97.4°W; 130 km, 8,500). They are is flanked by patches of fossae; **Tractus Fossae** (25.7°N, 101.4°W; 390 km) to the west, along with **Tractus Catena** (27.8°N, 102.7°W; 897 km) trend north south, while southeast of Uranius Mons the fossae and catena run at a variety of angles to each other.

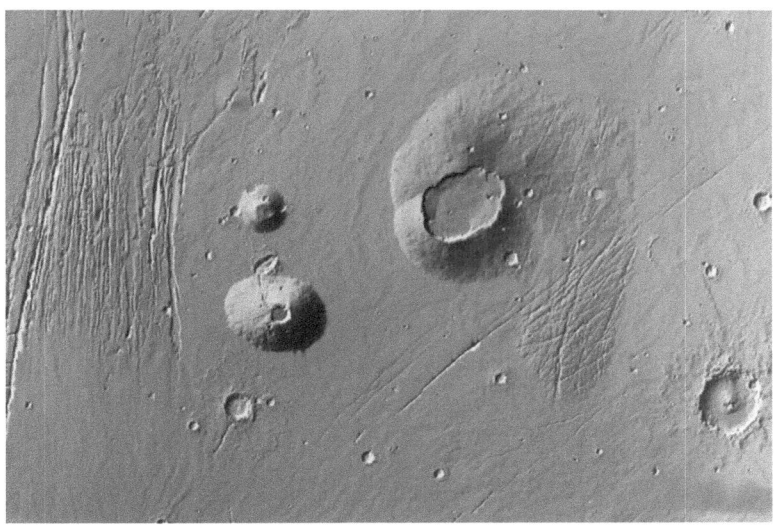

Uranius Mons and environs. Topographic image (Credit: NASA/JPL/Grego)

From a topographical perspective, the area north and west of Alba Mons, made up of Vastitas Borealis and the neighbouring lowland plains of Arcadia Planitia, are relatively bland and featureless from a large scale topographical point of view. The most sizeable feature in this area is the impact crater **Milankovic** (54.7°N, 146.7°W; 118 km), surrounded by the many low ridges of **Arcadia Dorsa** (54.7°N, 140.0°W; 1,900 km) which appear to curve in arcs parallel to the outline of the Alba uplift. Groups of dozens of small individual peaks make up the **Erebus Montes** (35.7°N, 175.0°W; 785 km) in western Arcadia Planitia; associated peaks also run south in western Amazonis Planitia.

Amazonis Planitia averages 3 km below datum. Its southern section around the equator is intruded upon by a number of broad southeast-northwest uplands, ridges and Mensae, including **Eumenides Dorsum** (4.5°N, 156.5°W; 566 km), **Amazonis Mensa** (2.0°S, 147.5°W; 500 km) and **Gordii Dorsum** (4.5°N, 144.1°W; 494 km); between these higher areas run outwash channels originating further south. Notable among these is **Mangala Valles** (11.5°S, 151.0°W; 828 km) whose course is sometimes sinuous and anastomosing and whose origin lies buried in the highlands of southwestern Tharsis and Daedalia Planum. Numerous parallel linear rift valleys cut across Daedalia Planum and the mountainous cratered highlands to its west; prominent among these is **Mangala Fossa** (17.3°S, 146.0°W; 688 km) and the very extensive **Sirenum Fossae** (34.6°S, 160.9°W; 2,735 km).

A series of mountain ridges marks the eastern border of the cratered landscape of Terra Sirenum. Of the more prominent craters in this region are **Newton** (40.8°S, 158.1°W; 298 km) and **Copernicus** (49.2°S, 169.2°W; 294 km). Newton, the younger of the pair, is impressed into the broad curving heights of central Terra Sirenum and has one of the deepest floors of all the craters in this region. Copernicus is in a considerably more eroded state and appears to have been deformed by crustal compression, a force that has produced the extensive north–south ridge upon which it sits.

Craters Newton (upper right) and Copernicus (lower left). Topographic image (Credit: NASA/JPL/Grego)

6.5 Region 3: Utopia-Elysium-Cimmeria Region (180–270°W)

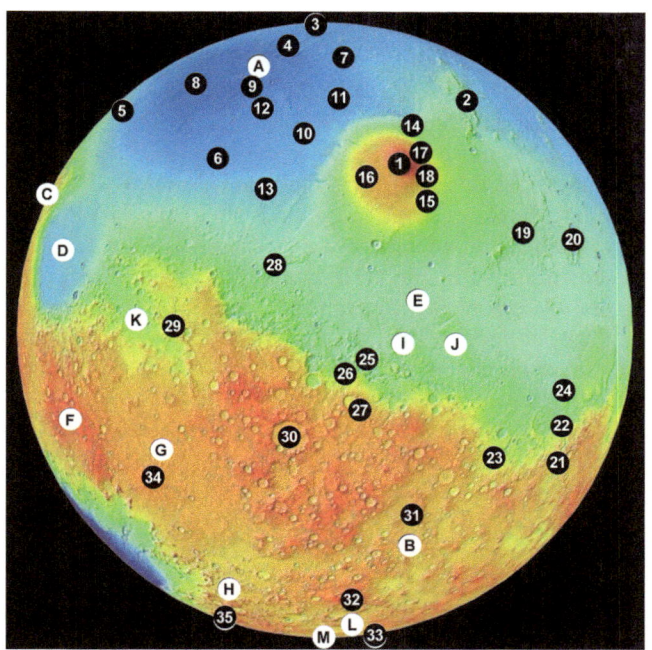

Map of Region 3, centred on the equator at 225°W, showing features mentioned in the descriptive text. Key (in order of first mention in the survey): A, Utopia Planitia; B, Cimmeria Terra; 1, Elysium Mons; C, Syrtis Major Planum; D, Isidis Planitia; 2, Phlegra Montes; 3, Panchaia Rupes; 4, Cydnus Rupes, Vivero; 5, Utopia Rupes; 6, Hephaestus Rupes; 7, Mie; 8, Nier; 9, Kufra; 10, Granicus Valles; 11 Hrad Vallis; 12, Tinjar Vallis; 13, Hebrus Valles, Hephaestus Fossae; E, Elysium Planitia; 14, Hecates Tholus; 15, Albor Tholus; 16, Elysium Chasma; 17, Stygis Fossae; 18, Stygis Catena; 19, Tartarus Montes; 20, Orcus Patera; F, Tyrrhena Terra; G, Hesperia Planum; H, Promethei Terra; 21, Ma'adim Vallis; 22, Gusev; 23, Al-Qahira Vallis; 24, Apollinaris Mons; I, Aeolis Planum; J, Zephyria Planum; 25, Aeolis Mensae; 26, Gale; 27, Wien; 28, Hyblaeus Dorsa; K, Amenthes Planum; 29, Amenthes Rupes; 30, Herschel; 31, Molesworth; 32, Kepler; L, Chronium Planum, Eridania Scopulus, Thyles Rupes; 33, Ulyxis Rupes; 34, Tyrrhenus Mons; 35, Secchi; M, Promethei Planum, Promethei Rupes. Note that letters refer to features of extended area, ie., terrae, planitiae, chaoses, paludes and plana, while numbers refer to all other features (Credit: NASA/Google Earth/Grego)

In broad topographical terms, this region of Mars is the easiest of the four to describe. The most obvious thing to notice is the clear-cut dichotomy between the north and south hemispheres. The relatively smooth, crater-free lowland plains of the north, largely occupied by **Utopia Planitia** (49.7°N, 242.0°W; 3,200 km) are sharply separated from the cratered highlands of **Cimmeria Terra** (33.0°S, 212.3°W; 5,856 km) in the south; the volcanic uplift upon which sits **Elysium Mons** (25.3°N, 212.8°W; 401 km) forms an island of high ground in mid-northern latitudes.

Although it's by no means an obvious fact, Utopia Planitia marks the site of Mars' biggest and oldest recognized impact basin. Three thousand and two hundred kilometers across, the basin's once mountainous outer ring can be seen in the southwest, where the ground rises towards northern **Syrtis Major Planum** (8.4°N, 290.50°W; 1,350 km) and **Isidis Planitia** (13.9°N, 271.6°W; 1,225 km), in the south towards Cimmeria Terra and most prominently in the east where it is bordered by

Phlegra Montes (40.4°N, 196.3°W; 1,351 km). Note that both Isidis Planitia and Syrtis Major Planum are discussed more fully in Region 4 below. Concentric arcs of ridges and scarps are also evident in the north, including **Panchaia Rupes** (63.7°N, 240.0°W; 1,500 km). Numerous long ridges and scarps radial to a smooth west–east elongated central area – presumably outlining the inner ring of the Utopia basin – can also be traced. Among these is **Cydnus Rupes** (55.7°N, 244.0°W; 1,600 km) to the north and **Utopia Rupes** (39.7°N, 270.0°W; 2,550 km) in the west. Radial ridges are also to be found in a broad fringe towards the southern edge of the Utopia basin, with the **Hephaestus Rupes** (21.8°N, 243.0°W; 1,750 km). The eastern part of the Isidis basin, formed after the Utopia basin, overlies southwestern Utopia (discussion of this is left for Region 4, below).

The Utopia impact has been dated to the early Noachian Period, but its low lying central regions were soon filled with lava flows some 2–3 km deep; major volcanism inside the basin ended in the early Hesperian Period. The sizeable volcanic complex at Elysium was probably prompted by the Utopia impact, where volcanic activity took place for longer, into the early Amazonian Period. Deposition by sediments from large bodies of water fed by drainage channels in the south provided additional infilling.

Mie (48.2°N, 220.4°W; 104 km) is the largest impact crater within Utopia Planitia, with a clear-cut terraced wall and knobbly radial ejecta; the area to its west contains a plethora of small named craters which would have remained unnamed had it not been the site of the Viking 2 lander (a fact of nomenclature that's reflected around other Mars landing sites). Lobate ejecta is evident around many of the craters in Utopia, notably those in a 'cluster' of large craters in central Utopia, including **Vivero** (49.0°N, 241.2°W; 28 km), **Nier** (42.8°N, 254.0°W; 47 km) and **Kufra** (40.4°N, 239.7°W; 38 km).

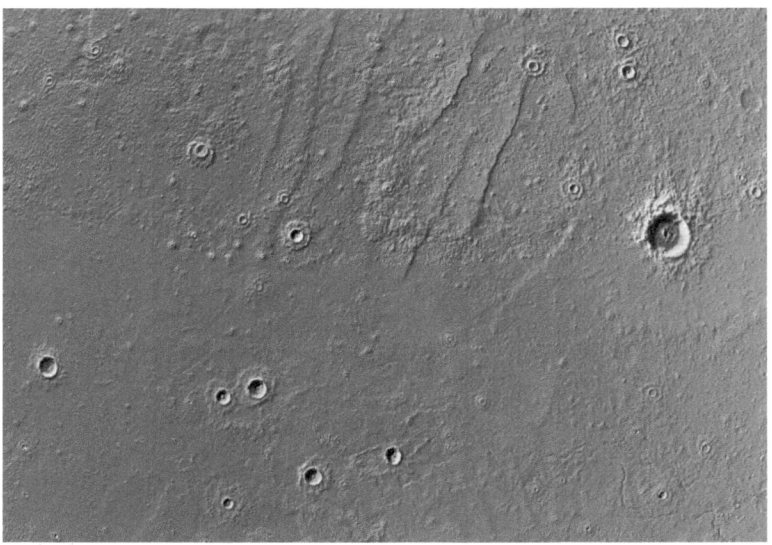

The central portion of Utopia Planitia, with the crater Mie (at right) and Cydnus Rupes (top half). Note the lobate ejecta around all the smaller craters. Topographic image (Credit: NASA/JPL/Grego)

A number of outflow channels and valley systems can be found in and around Utopia Planitia. Valley networks are concentrated an broad arc in northwestern Utopia, while a curving swathe of outflow channels in the northeastern central region includes **Granicus Valles** (29.7°N, 229.0°W; 750 km) which runs off the Elysium uplift, **Hrad Vallis** (38.4°N, 224.7°W; 825 km), **Tinjar Vallis** (37.7°N, 235.8°W; 425 km), **Hebrus Valles** (20.0°N, 234.0°W; 317 km) and **Hephaestus Fossae** (20.9°N, 237.5°W; 604 km).

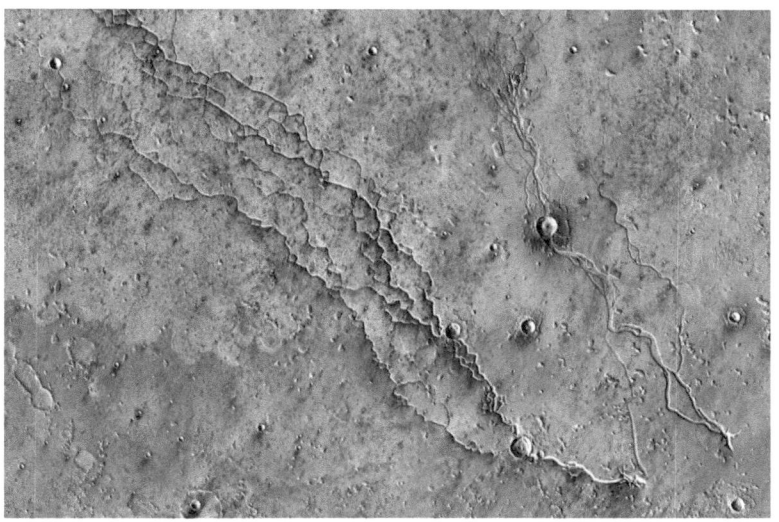

The Hephaestus Fossae valley network, with the smaller Hebrus Valles (at right). Topographic image (Credit: NASA/JPL/Grego)

Measuring around 1,600 km across, the domed volcanic uplift around Elysium Mons is Mars' second largest volcanic region. It's often claimed that the uplift lies on central **Elysium Planitia** (3.0°N, 205.3°W; 3,000 km), but in fact the IAU recognised extent of Elysium Planitia lies in a swathe well south of it, in the Cerberus region. Volcanism began here with the huge impact that carved out the Utopia basin, and it may have also been triggered further by the later Argyre and Hellas impacts. The Elysium dome has smaller volcanoes than Tharsis (5,500 km to its east); the main ones are Elysium Mons (13,860 m above datum), **Hecates Tholus** (32.1°N, 209.8°W; 183 km across, 4,720 m above datum) to its north and **Albor Tholus** (18.8°N, 209.6°W; 170 km, 4,500 m).

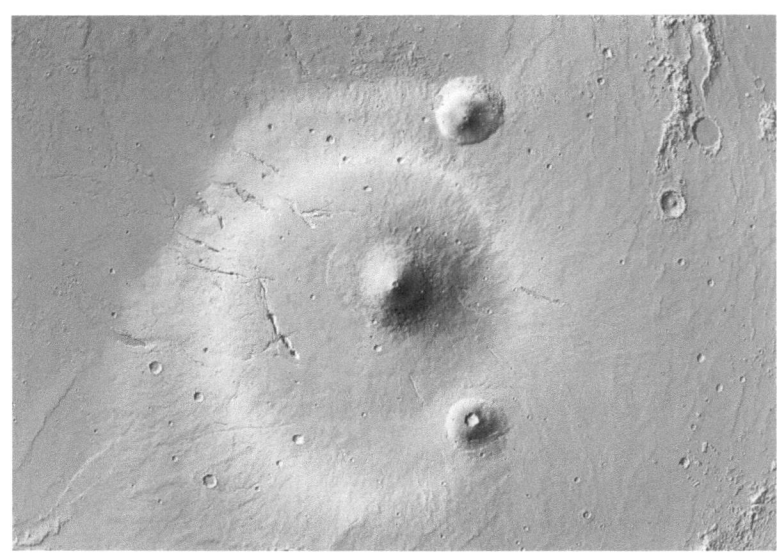

The volcanic region of Elysium, dominated by Elysium Mons (centre) with Hecates Tholus and Albor Tholus. Topographic image (Credit: NASA/JPL/Grego)

Elysium Mons is offset from the centre of the broad domed uplifted region from which it protrudes. The volcano is capped by a 15 km wide caldera, its steeper upper slopes crosses by lava flows and riven with scarps and fossae, including a steep arcuate cliff 125 km long some 170 km west of the volcano's center. A number of valleys and catenae lie further afield around its base, particularly in the east, in the vicinity of **Elysium Chasma** (22.3°N, 218.5°W; 129 km) and Granicus Valles, where they drain into the lowland plains of southeastern Utopia Planitia. **Stygis Fossae** (26.6°N, 210.3°W; 370 km) describe near-perfect concentric arcs some 190 km east of the center of Elysium Mons, which is taken up in the southeast by **Stygis Catena** (23.3°N, 209.5°W; 66 km).

Close-up of Elysium Mons and its caldera. Topographic image (Credit: NASA/JPL/Grego)

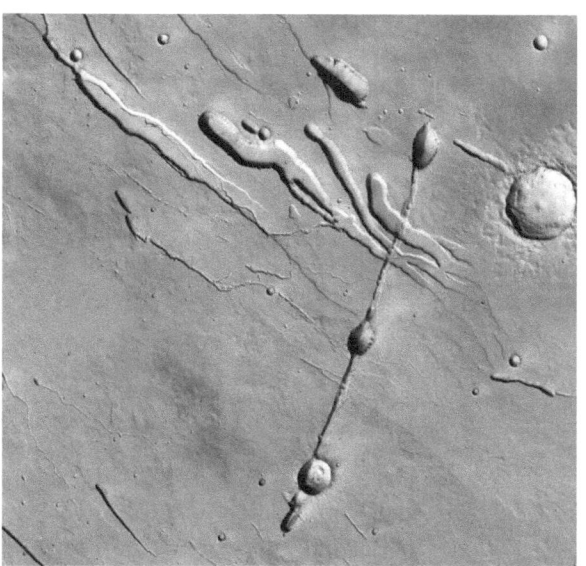

Stygis Catena (linear feature with collapse pits at both ends) cuts through a cluster of sinuous valleys on the eastern slopes of Eysium Mons. Topographic image (Credit: NASA/JPL/Grego)

Hecates Tholus lies just northeast of Elysium Mons, their bases separated by 60 km. It is capped by a caldera 10 km across at its widest (northeast-southwest), comprising multiple overlapping craters, the smallest and most recent of which is 7 km across with a floor 600 m below its southern rim. Several narrow valleys (flow features) and volcanic catenae surround the catena, and a well-preserved ash deposit west of the caldera is thought to have been laid down during an explosive eruption around 350 million years ago. A scalloped, flat-based elongated crater, 42 km long (northeast-southwest) overlaps the northwestern base of Hecates Tholus, possibly a collapse feature or an explosive crater whose base has been filled with lava, aeolian and fluvial deposits.

Albor Tholus, the smallest of the Elysium volcanic trio, lies at the extreme south-eastern base of Elysium Mons. It has a clearly-defined, east–west elongated ovoid base and its 3,000 m deep summit caldera measures 30 km across. Like the others, fossae formed by crustal tension and catenae, in some cases were material has fallen into vacated lava tubes, are to be found on its slopes.

Moving some 1,800 km east of Elysium Mons, beyond the scattered peaks of **Tartarus Montes** (16.0°N, 193.0°W; 1,070 km, 940 m below datum), lies the remark-able **Orcus Patera** (14.2°N, 181.5°W; 375 × 158 km), an elongated, north–south oriented crater. Of uncertain origin, it may be a conjoined or low-angle impact scar or (more likely) a volcanic collapse feature. Several east–west rilles, formed after Orcus Patera by tension as the crust was uplifted regionally, can be found either side of the feature, and one can be traced all the way across its floor.

The cratered southern highlands in Region 3 are more varied than a first glance may suggest. The main highland area, Cimmeria Terra, borders Sirenum Terra at 180°W (an arbitrary boundary undetermined by topography), runs adjacent to Elysium Planitia in the north; its western edge meets (from north to south) Isidis Planitia, **Tyrrhena Terra** (11.9°S, 271.6°W; 2,470 km), **Hesperia Planum** (21.4°S,

250.1; 1,601 km) and **Promethei Terra** (64.4°S, 263°W; 3,244 km). Numerous great outwash valleys run south from Cimmeria Terra to the lowland plains to its north. The most easterly of these, **Ma'adim Vallis** (21.6°S, 182.7°W; 825 km) originates in the highlands at the junction of several smaller valleys and runs a gently winding path, cutting through a number of older craters and finally drains into the crater **Gusev** (14.5°S, 184.6°W; 166 km). Further west, **Al-Qahira Vallis** (18.2°S, 197.5°W; 555 km) drains into the Elysium Planum lowlands. Three hundred kilometers north of Gusev lies **Apollinaris Mons** (9.3°S, 185.6°W; 296 km, 5,000 m), a small shield volcano with an irregular 68 km diameter caldera from which emanates a channel which spreads out on the volcano's southern flanks into a fan-shaped 'delta'. Apollinaris Mons' base has and an irregular, scalloped border, most pronounced in the west where it is marked by a steep cliff.

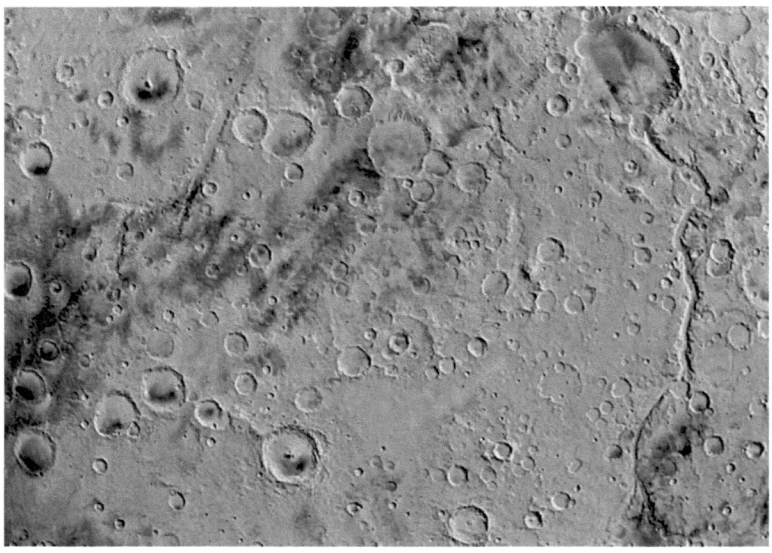

Outflow valleys Madim Vallis and the crater Gusev (at right) and Al-Qahira Vallis (at left) (Credit: NASA/JPL/Grego)

Aeolis Planum (0.8°S, 215.0°W; 820 km) and **Zephyria Planum** (1.0°S, 206.9°W; 550 km), both districts of southwestern Elysium Planitia, are crossed by networks of fluvial channels. The most prominent among these define the **Aeolis Mensae** (2.9°S, 219.6°W; 820 km), east of the prominent impact crater **Gale** (5.4°S, 222.3°W; 155 km). Yet more sizeable outflow channels cut the higher ground further south, notably those originating from the crater **Wien** (10.7°S, 220.4°W; 120 km) and its environs.

A system of long, broad ridges; one system, **Hyblaeus Dorsa** (10.9°N, 231.0°W; 875 km) crosses like a causeway from the hilly region adjacent to Cimmeria Terra over Elysium Planitia to the lower slopes of the Elysium uplift.

Northwestern Cimmeria Terra extends like a peninsula north of Hesperia Planum and Tyrrhena Terra, buffered by the broad (average width 250 km) lowland 'valley' of **Amenthes Planum** (3.2°N, 254.3°W; 960 km) whose northern edge is marked by the cliff **Amenthes Rupes** (1.6°N, 249.5°W; 331 km). Outflow channels continue to be found emanating from the cratered heights into the lowlands of southern Utopia Planita.

Larger craters of note in Cimmeria Terra include **Herschel** (14.7°S, 230.3°W; 304.5 km) whose floor displays the remnants of an inner mountain ring, **Molesworth** (17.4°S, 210.9°W; 181 km) whose inner mountain ring is only partially visible in the south, and **Kepler** (46.8°S, 219.1°W; 233 km) with a well-preserved inner ring 116 km across. Several large northeast-southwest ridges and cliffs can be found in Kepler's vicinity, around the plateau **Chronium Planum** (59.7°S, 220.0°W; 550 km), including **Eridania Scopulus** (53.5°S, 221.0°W; 939 km) and **Ulyxis Rupes** (68.4°S, 200.1°W; 390 km).

Rising above Hesperia Planum, a relatively young high plain to the west of Cimmeria Terra, is the unusual volcano **Tyrrhenus Mons** (21.4°S, 253.5°W; 473 km). From a 12 km diameter central crater, a broad smooth-floored valley opens up its southwestern rim and proceeds down its southwestern slopes. Pronounced ridges and valleys cut into the volcano's sides, where in places they are themselves cut across by arcuate valleys and faults. The mountain is thought to have been built up in layers of viscous lava flows and ash by successive Plinian-type eruptions, hence its dissimilarity with volcanoes in the provinces of Tharsis and Elysium, for example.

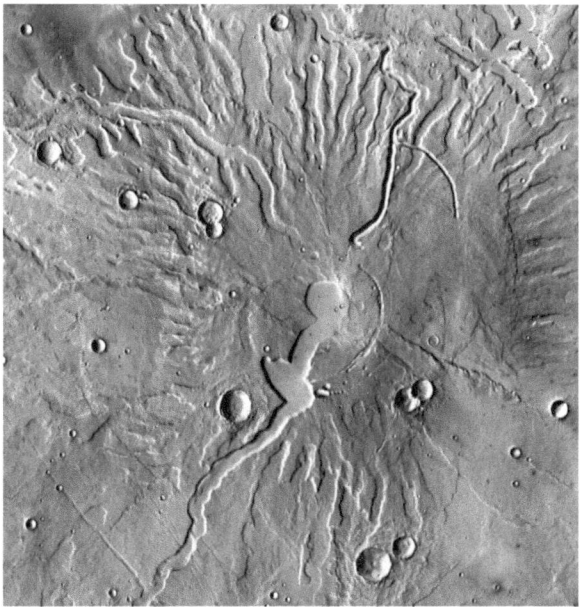

Tyrrhenus Mons, a Plinian-type volcano. Topographic image (Credit: NASA/JPL/Grego)

Promethei Terra, south of Hesperia Terra, borders the eastern ring of the mighty Hellas impact basin (discussed in Region 4 below). The largest crater in this heavily cratered area is **Secchi** (58.0°S, 258.1°W; 234 km) whose western floor contains a remnant arc of its inner mountain ring. Promethei Terra encroaches upon the Martian Antarctic Circle, where it contains numerous large ridges and escarpments, the most extensive of which are **Thyles Rupes** (69.6°S, 229.1°W; 675 km) and the vast south pole-concentric **Promethei Rupes** (75.3°S, 269.4°W; 1,248 km); the latter borders **Promethei Planum** (78.9°S, 270.0°W; 850 km), relatively smooth lowlands that border the south polar area.

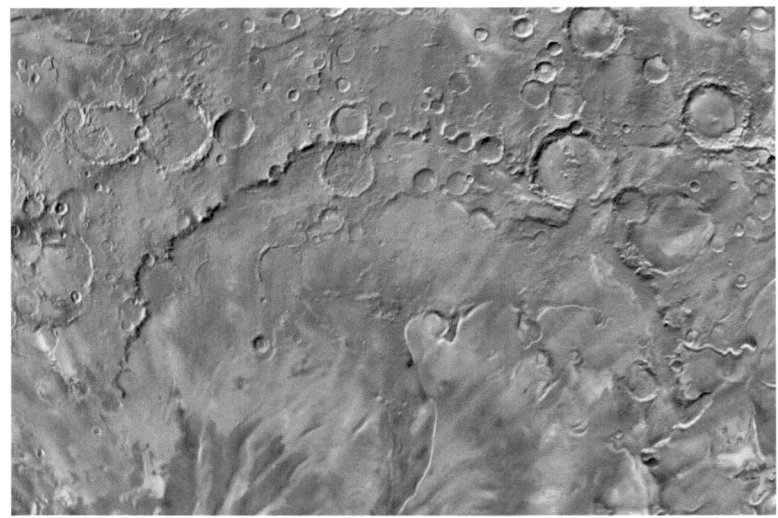

The near-south polar region, showing Promethei Rupes and Promethei Planum. Topographic image (Credit: NASA/JPL/Grego)

6.6 Region 4: Vastitas-Sabaea-Hellas Region (270–360°W)

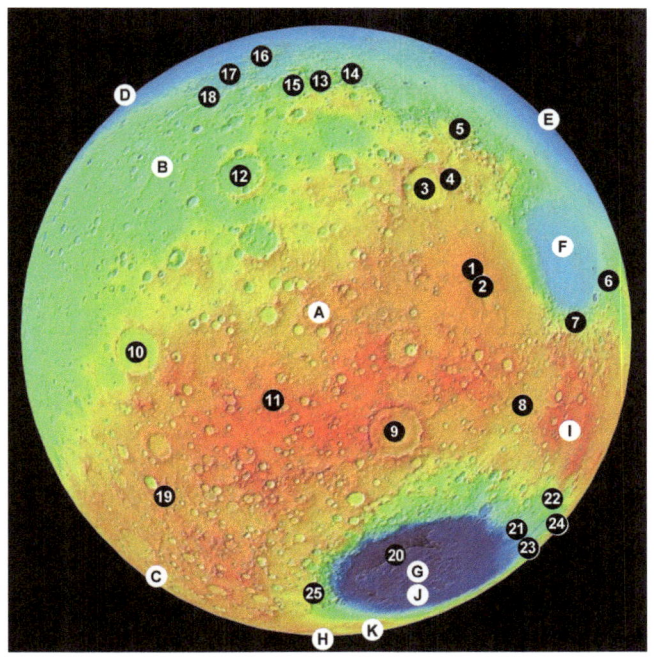

Map of Region Four, centred on the equator at 315°W, showing features mentioned in the descriptive text. Key (in order of first mention in the survey): A, Sabaea Terra; B, Arabia Terra; C, Noachis Terra; D, Acidalia Planitia; E, Utopia Planitia; F, Isidis Planitia, Isidis Dorsa; G, Hellas Planitia; H, Australe Planum; 1, Nili Patera; 2, Meroe Patera; 3, Antoniadi; 4, Baldet; 5, Nilosyrtis Mensae; 6, Amenthes Fossae; 7, Libya Montes; I, Tyrrhena Terra; 8, Oenotria Scopulus; 9, Huygens; 10, Schiaparelli; 11, Dawes; 12, Cassini; 13, Moreux; 14, Protinilus Mensae; 15, Ismeniae Fossae; 16, Lyot; 17, Deuteronilus Mensae; 18, Mamers Valles; 19, Scylla Scopulus; 20, Alpheus Colles; J, Hellas Chaos; 21, Dao Vallis; 22, Hadriacus Mons; 23, Harmakhis Vallis; 24, Centauri Montes; 25, Hellespontus Montes; K, Malea Planum. Note that letters refer to features of extended area, ie., terrae, planitiae, chaoses, paludes and plana, while numbers refer to all other features (Credit: NASA/Google Earth/Grego)

Most of Region 4 is covered with cratered uplands, central to which is the sprawling **Sabaea Terra** (2.8°N, 308.7°W; 4,688 km) with **Arabia Terra** (21.3°S, 354.3°W; 4,852 km) to its north, **Noachis Terra** (50.4°S, 5.2°W; 5,519 km) to its southwest and Tyrrhena Terra to its east. Four major impact basins, each containing lowland plains, encroach upon Region 4, namely **Acidalia Planitia** (46.7°N, 22.0°W; 2,300 km) in the northwest, **Utopia Planitia** in the northeast, **Isidis Planitia** in the east and **Hellas Planitia** (42.7°S, 290.0°W; 2,200 km) in the southeast. To the far north are the lowland plains of Vastitas Borealis, linking Utopia Planitia with Acidalia Planitia, while to the south lies Noachis Terra and **Australe Planum** (83.9°S, 200.0°W; 1,450 km) around the south pole.

Sabaea Terra occupies a huge area of this region, from 42°N to 37°S and 278°W to 351°W. Its northeastern district, Syrtis Major Planum, is a volcanic area, most of it noticeably less cratered than other areas of Sabaea Terra. Highest in the west, the area generally slopes to its lowest in the east where it has a well-defined boundary with Isidis Planitia. Because Syrtis Major Planum has gentle slopes averaging just

1°, it's less than obvious that Syrtis Major Planum is in fact a shield volcano – in terms of area, one of the largest on Mars, covering around 800,000 km². Central to Syrtis major Planum is a north–south elongated depression measuring 350 × 150 km whose edge is scalloped by the calderas **Nili Patera** (8.9°N, 293.0°W; 70 km) and **Meroe Patera** (6.9°N, 291.4°W; 50 km) the floors of which are around 2,000 m below their rims. A pattern of wrinkle ridges, caused by slumping and compression of the surface material after the magma chamber collapsed, can be found in and around the central depression and across the volcano's slopes in a general radial pattern. Additional arcuate ridges were formed by sliding and thrust faulting.

Syrtis major planum volcanic shield. Topographic image (Credit: NASA/JPL/Grego)

Immediately north of Syrtis major Planum lies the large crater **Antoniadi** (21.3°N, 299.2°W; 394 km) its northeastern rim overlapped by the younger crater **Baldet** (22.8°N, 294.6°W; 180 km). Both show signs of an inner ring structure. This far northern area of Sabaea Terra is heavily cratered, and its 'shoreline' with Vastitas Borealis is crossed by valleys – that extend to the hummocky plains of **Nili Colles** (38.7°N, 297.1°W; 645 km). The landscape is further broken up by fossae, creating a complex of smaller angular 'islands' including **Nilosyrtis Mensae** (34.7°N, 292.1°W; 705 km). Extensive broad arcuate fossae are also to be found east of Antoniadi on the northwestern margin of Isidis Planitia, a large impact basin that post-dates and overlaps the Utopia impact basin. Isidis Planitia is fairly smooth and bland, save for a smattering of small impact craters and a system of wrinkle ridges, the **Isidis Dorsa** (12.1°N, 271.8°W; 1,100 km). The Isidis plains open up to Utopia in the northeast, and only a suggestion of the now-buried main mountain ring can be seen here in the form of slightly higher, hilly terrain. Parallel arcuate rilles of **Amenthes Fossae** (8.7°N, 258.2°W; 889 km) can also be found on the other side of Isidis Planitia, and the basin's southern border incorporates the **Libya Montes** (2.8°N, 271.1°W; 1,170 km). These mountains define the northern edge of **Tyrrhena Terra**, an ancient volcanic province with a raised central portion; the area is heavily cratered. It is cut through by **Oenotria Scopulus** (11.0°S, 283.1°W; 1,360 km), a very large arcuate north-facing cliff which extends west into Sabaea

Terra all the way to the southern edge of Syrtis Major Planum. This huge escarpment is likely to be a crustal fault feature associated with the Isidis impact.

The southern half of Sabaea Terra is highly cratered. Its most prominent crater, **Huygens** (14.0°S, 304.4°W; 467 km), is Mars' largest impact crater; the second largest, **Schiaparelli** (2.8°S, 343.2°W; 458 km), lies on the western edge of Sabaea Terra, 2,300 km west of Huygens. Between the two, atop the highest terrain in the area, lies the crater **Dawes** (9.1°S, 321.9°W; 185 km). In terms of spectacle Huygens wins by virtue of its better state of preservation, with a broad inner ring system and a somewhat polygonal outer ring with steep internal slopes and an outer glacis cut through by radial gullies.

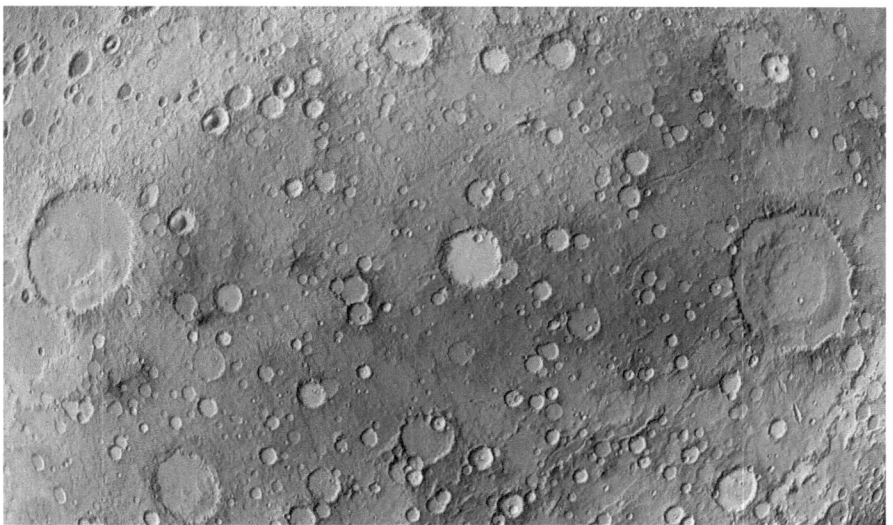

Central Sabaea Terra, home to the two largest craters on Mars – Schiaparelli (left) and Huygens (right). Crater Dawes lies at *centre*. Topographic image (Credit: NASA/JPL/Grego)

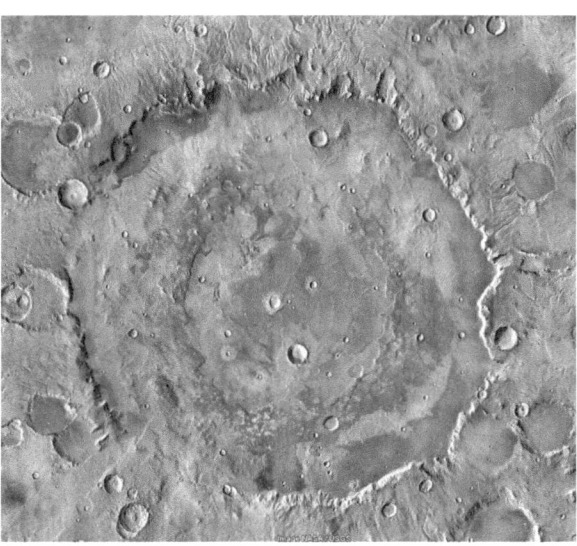

Huygens , Mars' largest impact crater (Credit: NASA/JPL/Grego)

Arabia Terra, one of the oldest surfaces on Mars, is occupies a wedge-shaped tract north of Sabaea Terra, and is on average some 2,000 m lower than its neighbour. The crater **Cassini** (23.4°N, 327.9°W; 408 km) lies in northeastern Arabia Terra, while further northeast **Moreux** (41.8°N, 315.6°W; 138 km) is located amid the eroded landscape of **Protinilus Mensae** (43.9°N, 310.6°W; 1,050 km) and **Ismeniae Fossae** (41.3°N, 322.3°W; 270 km). **Lyot** (50.5°N, 330.7°W; 236 km), a crater with a striking inner ring and radial ejecta, can be found on the adjoining plain, near the scattered, angular flat-topped 'islands' of **Deuteronilus Mensae** (43.6°N, 337.4°W; 937 km) which flank northern Arabia Terra. Like many of Mars' Mensae, they are essentially rock glaciers – water ice covered with a relatively thin layer of insulating regolith and shaped by aeolian processes. The area is encroached upon by numerous valleys originating in northern Arabia Terra, including **Mamers Valles** (40.0°N, 342.2°W; 1,020 km).

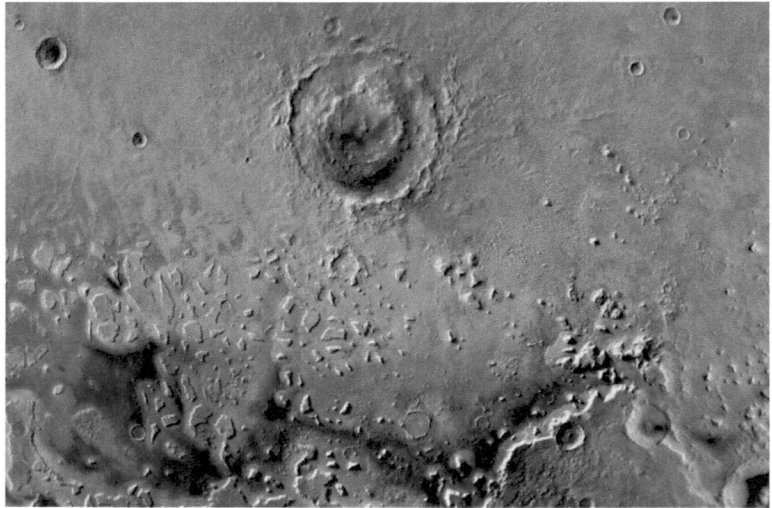

Lyot and Deuteronilus Mensae (Credit: NASA/JPL/Grego)

In a continuation of cratered highland terrain, southwestern Sabaea Terra joins northeastern Noachis Terra. The area, east of a very large unnamed crater (37.0°S, 356.0°W; 430 km) is crossed by several large, wide rift valleys bordered by steep scalloped cliffs, the largest of which is **Scylla Scopulus** (25.2°S, 341.7°W; 445 km). These features may be related to crustal forces imparted by the Hellas impact.

Hellas Planitia occupies the central lowlands of the Hellas impact basin. Measuring 2,300 km east–west and 1,800 km north–south, it is one of the largest impact scars in the Solar System; its formation took place in the Noachian Period of Mars, during the Solar System's Late Heavy Bombardment Period. Ovoid and east–west elongated, Hellas Planitia is centred west of a larger, more irregular outer ring.

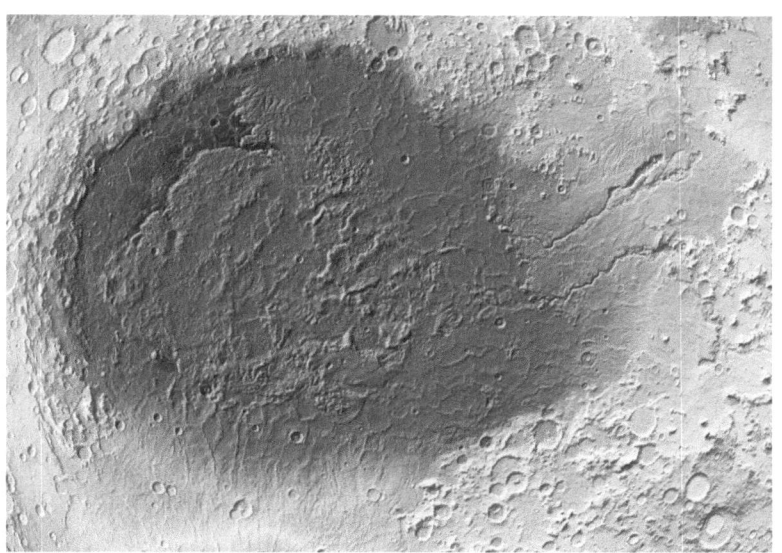

Hellas Planitia. Topographic image (Credit: NASA/JPL/Grego)

West of the center of Hellas Planitia spread the knobbly, ridged lowlands of **Alpheus Colles** (39.7°S, 300.3°W; 628 km), whose western border drops sharply to a deep, curving ridged plain (about 240 km wide) containing the lowest points on the planet, some 9,000 m below the level of the rim. To its south is the broken terrain of **Hellas Chaos** (47.5°S, 296.0°W; 595 km). A plain crossed by both radial and arcuate ridges curves around the north, east and southern parts of Hellas Planitia; this is fed into by large fluvial valleys from the higher ground to the east, including **Dao Vallis** (38.4°S, 272.1°W; 816 km) which originates in the southern foothills of the volcano **Hadriacus Mons** (32.1°S, 268.2°W; 575 km), and **Harmakhis Vallis** (40.5°S, 269.6°W; 475 km) whose origin is at the foot of **Centauri Montes** (38.6°S, 264.8°W; 270 km). Centauri Mons forms an eastern component of the irregular outer ring of the Hellas impact basin; the western component is more clearly defined by the curving sweep of **Hellespontus Montes** (44.4°S, 317.2°W; 730 km). More gently blending into the southern part of the Hellas basin is the upland plateau **Malea Planum** (64.8°S, 295.0°W; 900 km), a volcanic region containing several circular depressions surrounded by wrinkle ridges.

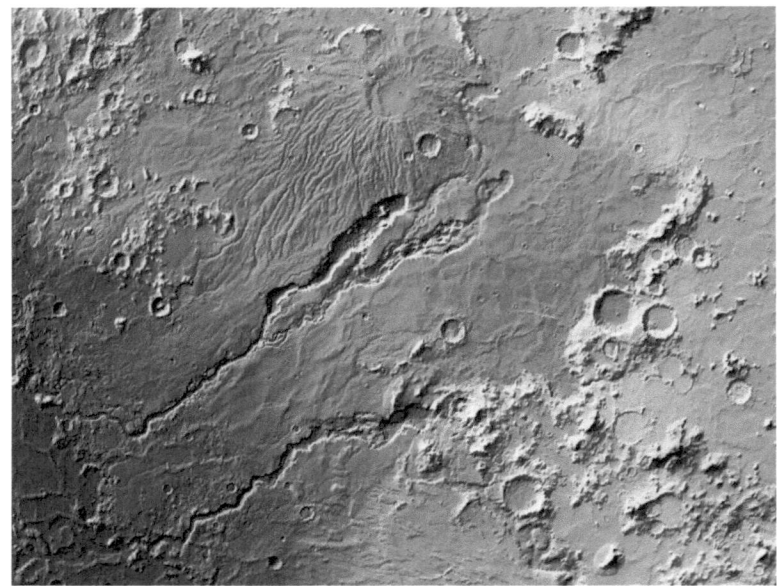

Eastern Hellas, showing (top to bottom) Hadriacus Mons, Dao Vallis, Harmakhis Vallis and Centauri Montes. Topographic image (Credit: NASA/JPL/ Grego)

Part II

Observing Mars

Chapter 7

Mars and How to Observe It

7.1 Introduction

Mars observation is among the most rewarding pursuits in amateur astronomy. Even though Mars is the Solar System's fourth planet and is half the Earth's diameter, much of the time it is so far away that its disk is reduced to a tiny orange ball less than 5 arcseconds across, shining sufficiently dimly to escape the immediate attention of all but vigilant sky watchers. But, for a few months every other year, Mars puts on a show that thrills visual observers and imagers. As its disk grows above 10 arcseconds across its features can be seen with ease and its brightness in the negative magnitudes screams out for attention.

At the climax of each apparition, when Mars is at its nearest during that showing and its magnitude and apparent diameter are at their greatest, amateur astronomers once again realize that their eager anticipation and expectations will be more than met. As the orange disk of Mars is centered in the field of view and focused, the observer can't help being struck by the fact that this is a world in many respects very much like our own, with icy polar caps, dusky 'seas' and brighter 'continents' swathed in an atmosphere capable of producing visible weather.

No two views of Mars are quite the same, and no two apparitions host an identical range or sequence of phenomena; this degree of unpredictability, in terms of what the planet will look like from one night to the next, during the course of an apparition and from one apparition to another, adds to the fascination of observing Mars.

Mars and
How to
Observe It

P. Grego, *Mars and How to Observe It*, Astronomers' Observing Guides,
DOI 10.1007/978-1-4614-2302-7_7, © Springer Science+Business Media New York 2012

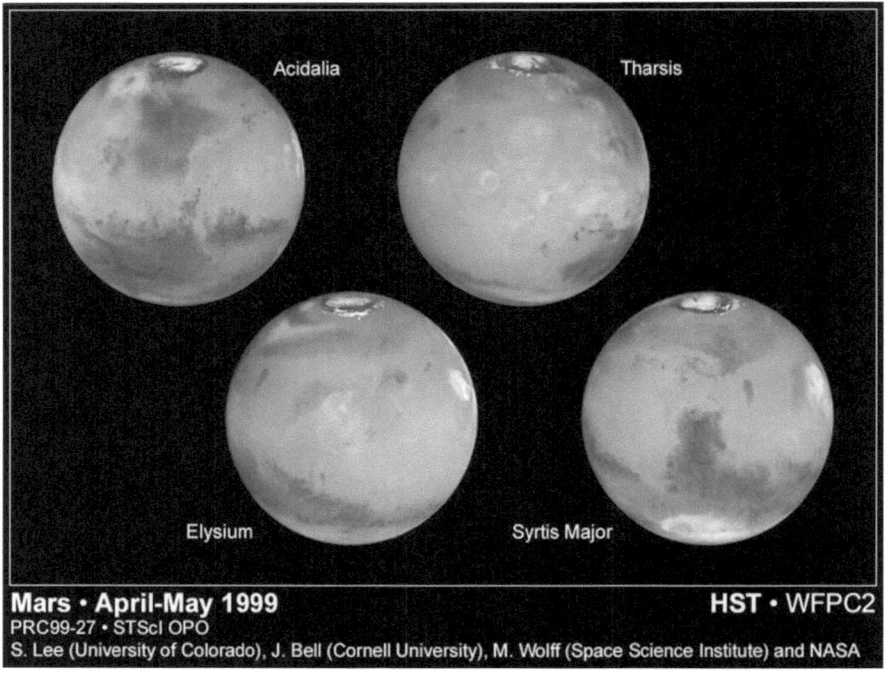

Mars, imaged by the Hubble Space Telescope during March and April 1999 (Credit: NASA/STScI)

Hubble Space Telescope view of Mars during its closest approach in late August 2003 (Credit: NASA/STScI)

7.2 Apparitions and Oppositions

Mars is called a superior planet because it orbits the Sun outside the orbit of Earth. Conjunction with the Sun takes place when Mars is on the far side of the Sun from Earth. Since Earth is on a faster orbital circuit than Mars, it appears to edge west of the Sun after conjunction. As it clears the Sun's glare, Mars begins to peek out of the pre-dawn skies. At this stage Mars has its smallest apparent diameter.

Mars only becomes a visually interesting object when it grows larger than 5 arcseconds in apparent diameter, when its ice caps and broader desert markings can just about be discerned through a 100 mm telescope. The magical 5 arcseconds apparent diameter is reached many months after Mars first becomes visible with the unaided eye in the morning skies.

At opposition, Mars is opposite the Sun and due south at midnight. It is virtually 100% illuminated at opposition and at its largest apparent diameter for that particular apparition. Since Mars has a markedly elliptical orbit, its distance from us varies at opposition. Mars displays a great variation in opposition diameter, ranging from a minimum of around 15 arcseconds at aphelic oppositions (when it is furthest from the Sun) to 25 arcseconds at perihelic oppositions (when it is nearest the Sun).

Mars' slow drift to the west of the Sun during the first part of its apparitions is caused by the movement of the Earth around the Sun. Although the aggregate movement of Mars against the celestial background is slowly eastwards, a phenomenon called retrograde motion causes it to reverse direction among the stars for a while, the planet performing a small loop or a zigzag in the sky before proceeding eastwards once more. Beginning some months prior Mars' opposition and ending some months afterwards, retrograde motion is caused by our view of the planet from the Earth and our shifting line of sight. Following conjunction, Mars appears to move sedately to the east. As the Earth, moving along its faster inner circuit, catches up with the planet prior to its opposition, our moving line of sight causes Mars' apparent motion to slow down and reverse direction, moving westwards for a while. As we draw further away from the planet after opposition, our line of sight begins to alter the apparent course of Mars; it appears to slow down and then finally recommences its slow eastward path.

Apparitions begin when the planet first becomes visible west of the Sun in the morning skies and ends when it finally vanishes into the glare of evening twilight. Mars is visible for around 18 months during each apparition, during the course of which its apparent diameter grows from a minimum of around 3.5 arcseconds to more than 14 arcseconds (at aphelic oppositions) and 25 arcseconds (at perihelic oppositions). Visual observations are possible throughout this period, although most observers consider that Mars is far too small for meaningful observation in the first few months at the beginning and end of each apparition, when the planet is less than 5 arcseconds across. Experienced CCD imagers are able to tease details out of much smaller Martian disks, so their apparition season extends beyond that of the visual observer.

Early or late in each apparition the broader, darker features are visible through modest sized telescopes, but the fine detail which makes Mars so fascinating to observe can only be discerned when Mars is close to opposition and the atmospheres of the Earth and Mars itself are clear. When Mars exceeds 5 arcseconds in apparent diameter, the larger dusky features and the bright north or south polar cap may at least be discerned through a 150 mm telescope. Many visual observers consider that only a Martian disk larger than 8 arcseconds is worthy of serious telescopic scrutiny, neglecting regular observation until Mars is well-established in the night skies, perhaps 6 months or even longer after its conjunction with the Sun.

Since Mars' orbital path is so eccentric, not all apparitions are equally favorable to observe. Mars' opposition cycle averages 15.8 years and involves five distant oppositions and two consecutive perihelic oppositions; the cycle repeats every 79 years or so. Oppositions occur when Mars is directly opposite the Sun around every other year, the average interval between oppositions being 780 days (around 50 days later in the year for each successive opposition). Since Mars is moving more slowly when further from the Sun, the interval between successive distant oppositions is actually less than that separating perihelic oppositions; for example, the 2012 and 2014 oppositions are 766 days apart, while 810 days separates the closer 2018 and 2020 oppositions.

At around opposition Mars is at its brightest and its apparent diameter is at its greatest that apparition. Oppositions don't occur at exactly the same time as closest approach; during the famous 2003 perihelic opposition Mars was closest to the Earth on 27 August, just a day prior to opposition. During the following apparition closest approach occurred on 30 October 2005, 8 days before opposition. The difference between opposition date and closest approach to Earth can be as much as a fortnight.

Oppositions of Mars, 2012–2022

Date	Dia. (arcseconds)	Dist. (km)	Mag.	Dec. (Constellation)
3 March 2012	13.9	101,106	−1.1	10° 10′ (Leo)
8 April 2014	15.1	93,106	−1.5	−05° 14′ (Virgo)
22 May 2016	18.4	76,106	−2.1	−21° 40′ (Scorpius)
26 July 2018	24.2	58,106	−2.8	−25° 23′ (Capricornus)
13 October 2020	22.4	63,106	−2.6	05° 30′ (Pisces)

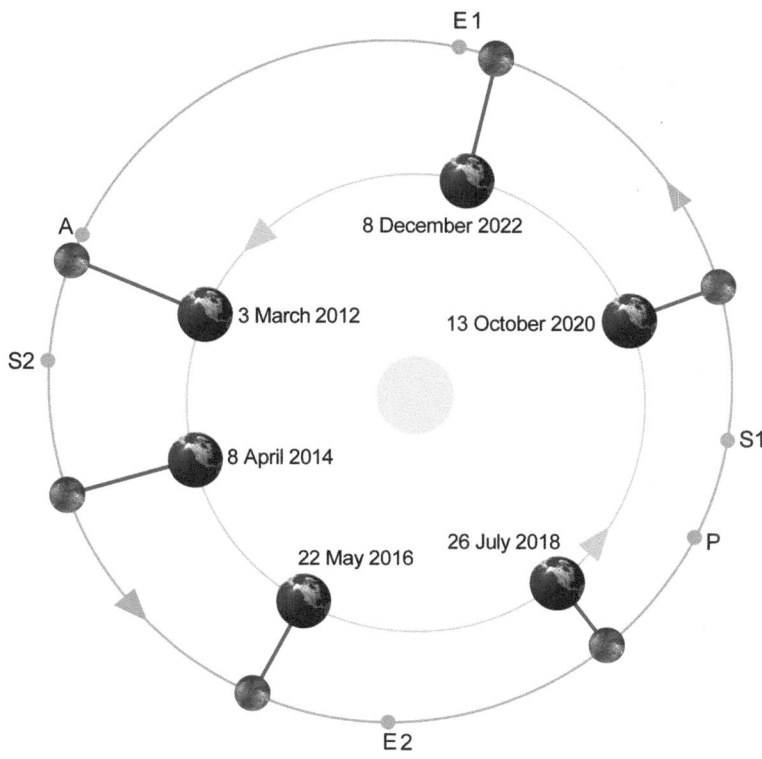

Oppositions of Mars, 2012–2022. E1 = Northern spring equinox, southern autumn equinox; E2 = Northern autumn equinox, southern spring equinox; S1 = Northern winter, southern summer solstice; S2 = Northern summer, southern winter solstice; P = Perihelion; A = Aphelion

Mars is further than 100 million kilometers from Earth at aphelic oppositions, presenting a disk less than 14 arcseconds in apparent diameter and shining at magnitude −1.1. Mars' north pole is always tilted towards the Sun during aphelic oppositions, and features in its brighter northern hemisphere are more favorably presented to us. At aphelic opposition, a telescopic magnification of 120x is required to make Mars appear the same apparent size as the full Moon viewed with the unaided eye (half a degree across).

Perihelic oppositions take place every 15–17 years, when Mars approaches closer than 55 million kilometers – around 150 times the distance from the Earth to the Moon. On these occasions, it presents a fully illuminated disk around 25 arcseconds in apparent diameter and shines at a brilliant magnitude −2.8. Perihelic oppositions see Mars' south pole tilted towards the Earth, and the duskier, feature-packed southern hemisphere is presented favorably to us. Although the bright south polar ice cap appears prominent, regular observers will have noted that shrinks noticeably during the months leading to opposition; summer temperatures in Mars' southern hemisphere have been on the rise, causing the outer parts of the cap to retreat.

Comparison between the apparent size of Mars during perihelic opposition (2003, left) and aphelic opposition (1999) (Credit: Grego)

A good 150 mm telescope will comfortably resolve objects separated by 1 arcsecond; at perihelic opposition, when Mars is around 25 arcseconds across, 1 arcsecond is equivalent to about 280 km on the planet. This is about the same resolution as a naked eye view of the Moon. At perihelic opposition a magnification of just 72x will show Mars at the same apparent diameter as the full Moon viewed with the unaided eye.

7.3 Seasons

With its axial tilt of 25.2°, Mars experiences a cycle of seasons, just like the Earth, only they last around twice as long. Mars' changing tilt toward us alters the appearance (through perspective) of the well-known dusky features and the apparent position and extent of the planet's north and south polar ice caps. How the tilt of Mars presents itself to the terrestrial observer depends on where Mars lies in its orbit and where the Earth is in relation to Mars, and the most obvious indicator of the planet's tilt is produced by the planet's polar caps. A 150 mm telescope will usually reveal one or the other polar cap (or their cloudy hoods) at any time during an apparition when the planet is larger than 5 arcminutes in apparent diameter.

Close oppositions see summer for Mars' southern hemisphere; the planet's south pole is tilted to its maximum extent towards the Sun and the south polar cap is small but prominent; the equator curves steeply across the north of the planet and all regions above 65°N never rotate into view (they are in permanent darkness as it is mid-winter for the northern hemisphere). It is summer for the planet's northern hemisphere during distant oppositions, when the south pole is in winter darkness and the north pole is well presented to us.

In terrestrial days, the length of Martian seasons is: Southern spring, northern autumn, 146 days; southern summer, northern winter, 160 days; southern autumn, northern spring, 199 days; southern winter, northern summer, 182 days. Therefore, the planet's southern hemisphere has a shorter, warmer summer as it is nearer the Sun, causing the south polar cap to melt more quickly than does the north polar cap during its summer. However, southern winter is longer and more severe than in the north; the south polar cap can extend to 60°S during southern winter, while the northern cap extends to just 50° during northern winter.

7.4 Rotation

Mars rotates on its axis once every 24.6 h, and observational drawings separated by an hour or so will easily reveal the planet's rotation. Since a day on Mars is slightly longer than a day on Earth, it follows that if Mars is observed at about the same time each evening, Mars' rotation will appear to lag behind, with features more easterly on the surface coming into view. The difference amounts to around 9° of Martian longitude east per day, so it follows that a complete tour around the planet can be accomplished in around 6 weeks if an observation is made at the same time each clear evening.

Such observations made over a period of time will clearly show progressively more easterly features, giving the illusion that Mars were rotating in a retrograde fashion. For example, beginning with Sinus Meridiani (on the planet's zero degree line of longitude), Hellas and Syrtis Major will be on the meridian around a week later; Elysium and Mare Cimmerium around 9 days afterwards; Tharsis after a further 10 days, followed 3 days later by Solis Lacus, with Mare Acidalium appearing on the meridian around a week later.

10 August 2003   15 August 2003   20 August 2003   25 August 2003   4 September 2003

A sequence of CCD images showing different aspects of Mars at 5 day intervals around the time of its close opposition in August 2003 (Credit: Jamie Cooper)

Chapter 8

A Tour of Mars Through the Eyepiece

8.1 A Tour of Mars, Westwards from 0° to 360°W, in Four Sections

We have the nineteenth century astronomers Wilhelm Beer and Johann Mädler to thank for establishing the prime meridian of Mars. They defined a permanent reference point on the planet's surface which was used for determining an accurate value for the planet's rotation and to construct accurate maps of Mars. This reference point, centered on a small feature we now know as Sinus Meridiani (Meridiani Planum), is still used today as zero longitude.

In order to be consistent with the system of longitude used by visual observers, and for ease of reference, all longitude co-ordinates in this book are given west of the Martian prime meridian; where possible, these have been taken from the IAU official list of Martian albedo features (determined in 1958) but where unnamed in this list, feature names and co-ordinates have been taken directly from Eugène Antoniadi's detailed map (from *La Planète Mars*, 1930). As with the topographic maps featured in the first part of this book, there is some overlap in description at the boundaries of each region.

Many of the descriptions featured in this survey of telescopically observable features of Mars have been based on my own telescopic observations made during numerous apparitions of Mars since 1982. I have also consulted Antoniadi's superb maps made around a century ago, which were based on his own meticulous observations, since these remain unequalled in terms of clarity and accuracy. Antoniadi's nomenclature has been used throughout. The observational illustrations are derived from more recent apparitions of Mars, with my own observational drawings (in pencil, PC enhanced and PDA drawing media) and also the excellent pencil work of Paul Stephens.

Most observing literature portrays Mars with south at top, the classical telescopic view. Nowadays many observers use a refractor, Maksutov-Cassegrain (MCT) or Schmidt-Cassegrain (SCT) telescope in conjunction with a star diagonal; this accessory makes such instruments less awkward to look through when pointed higher in the sky. However, mirror diagonals (when used in the northern hemisphere) show north towards the top and give an east–west flipped image,

P. Grego, *Mars and How to Observe It*, Astronomers' Observing Guides,
DOI 10.1007/978-1-4614-2302-7_8, © Springer Science+Business Media New York 2012

while prism diagonals rotate Mars by 180°, showing south at top but without flipping the image east–west.

To help the observer, I have presented three basic albedo maps of Mars, labeled and unlabeled. The first has north at top, celestial west at right; this matches the orientation of the topographic maps in the first part of this book, and tallies with the view for users of prism diagonals in the northern hemisphere. The second has north at top, celestial west at left (an east–west reversed view for users of mirror diagonals), while the third shows south at top, celestial west at left (the classical telescopic view).

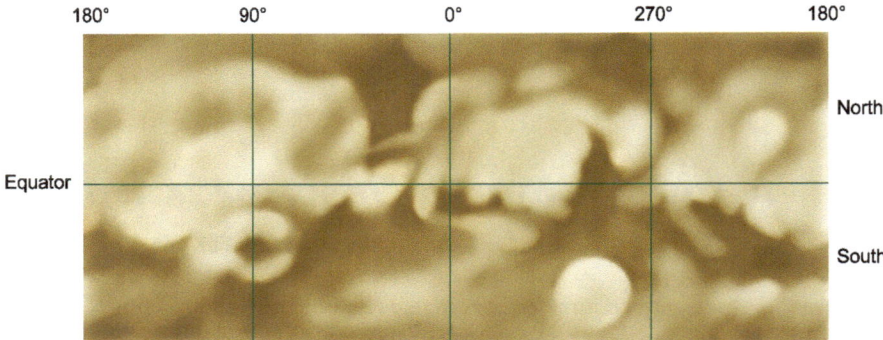

Map of Mars, showing albedo features. North at top, celestial west at right (Credit: Grego)

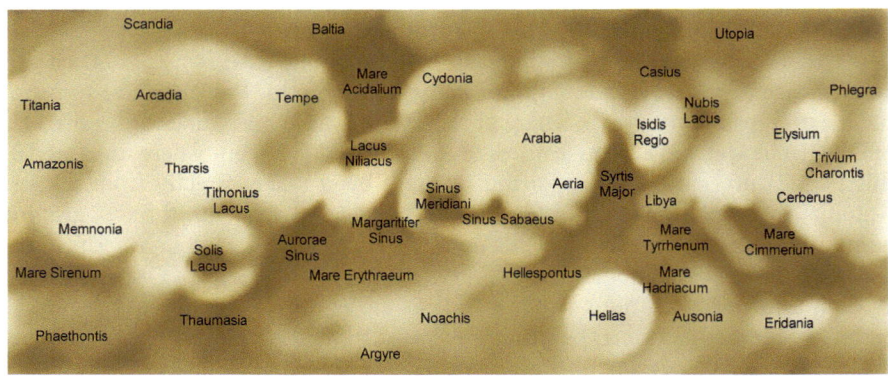

Labeled Map of Mars, showing albedo features. North at top, celestial west at right (Credit: Grego)

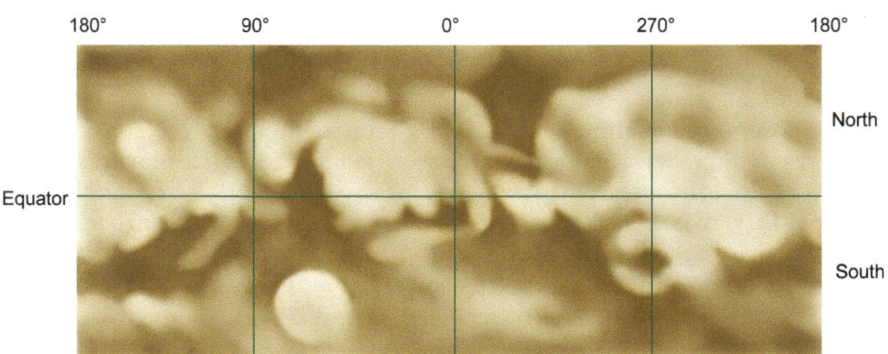

Map of Mars, showing albedo features. North at top, celestial west at left (Credit: Grego)

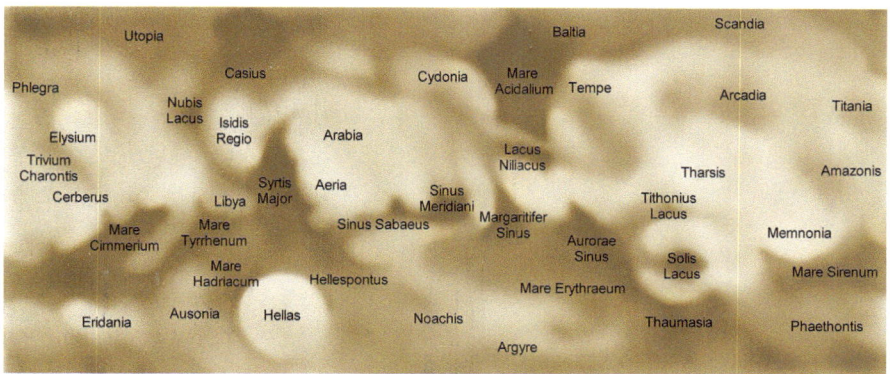

Labeled Map of Mars, showing albedo features. North at top, celestial west at left (Credit: Grego)

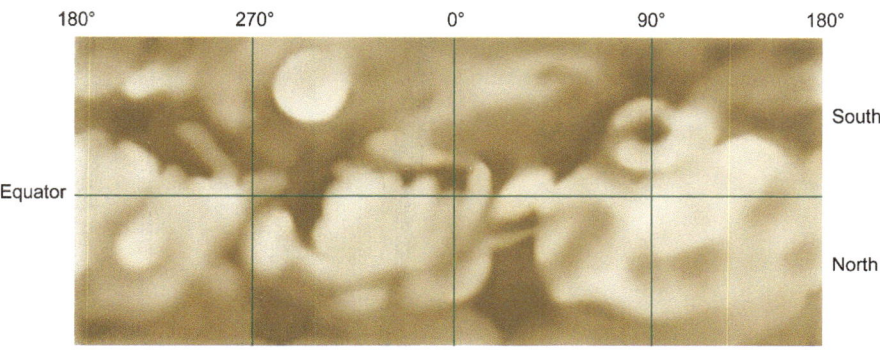

Map of Mars, showing albedo features. South at top, celestial west at left (Credit: Grego)

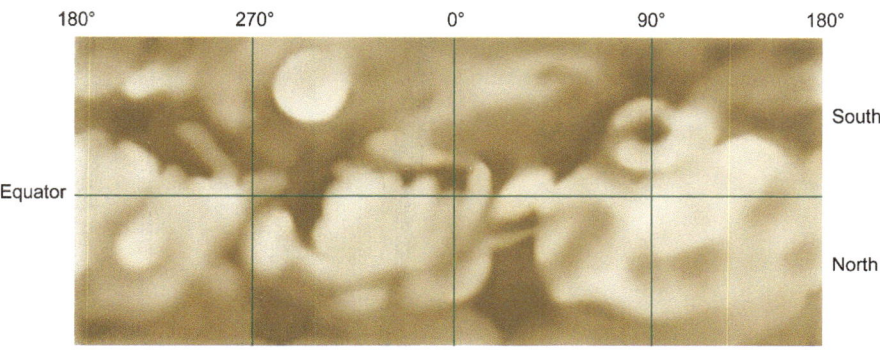

Labeled Map of Mars, showing albedo features. South at top, celestial west at left (Credit: Grego)

The hemispheric views accompanying each region below are presented with north at top, with Martian geographical west increasing to the left; these views match the topographic maps presented in the first part of this book. These views show albedo features and are based on Viking orbiter images taken in the late 1970s. However, it must be noted that Mars' albedo features can appear to vary in both outline and intensity during the course of an apparition and from one apparition to the next, owing to Martian weather.

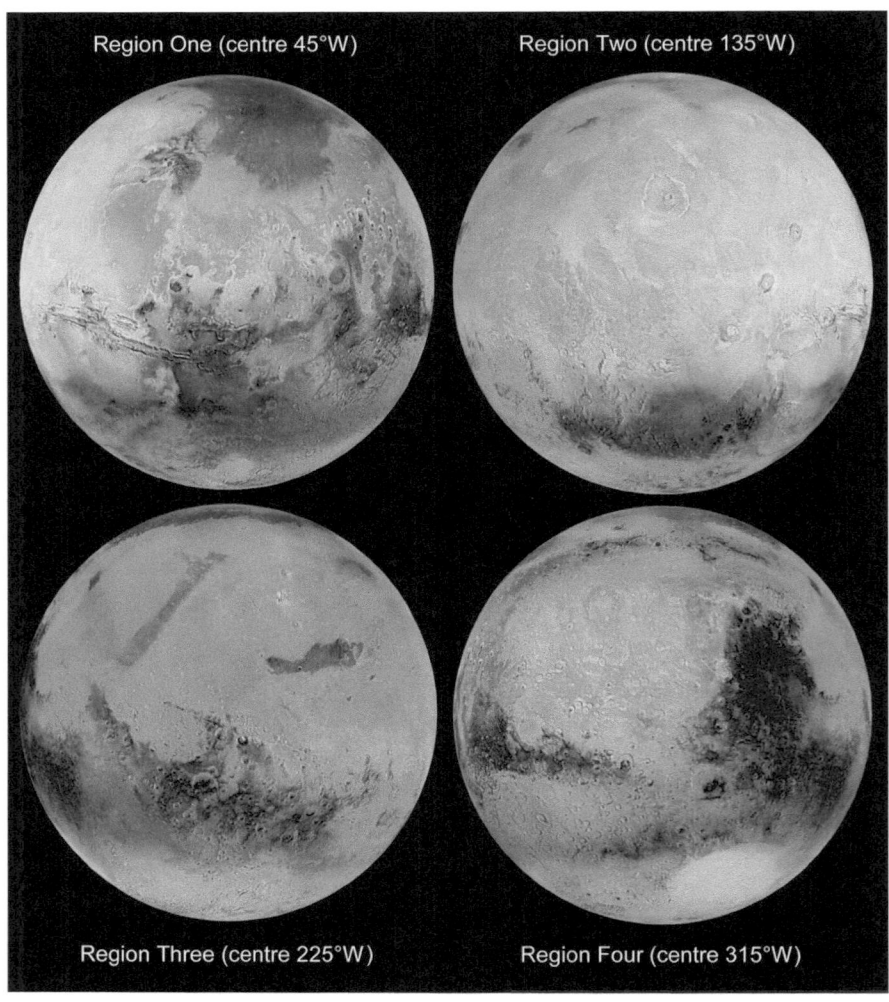

Region One (centre 45°W) Region Two (centre 135°W)

Region Three (centre 225°W) Region Four (centre 315°W)

Viking-based hemispheric albedo maps showing the four regions of Mars (north at top) (Credit: NASA/Google Earth/Grego)

Following each hemispheric view there is a more detailed albedo map of each region, based upon Antoniadi's map of Mars. These maps, all with north at top, show all the features mentioned in the text below, plus numerous additional features.

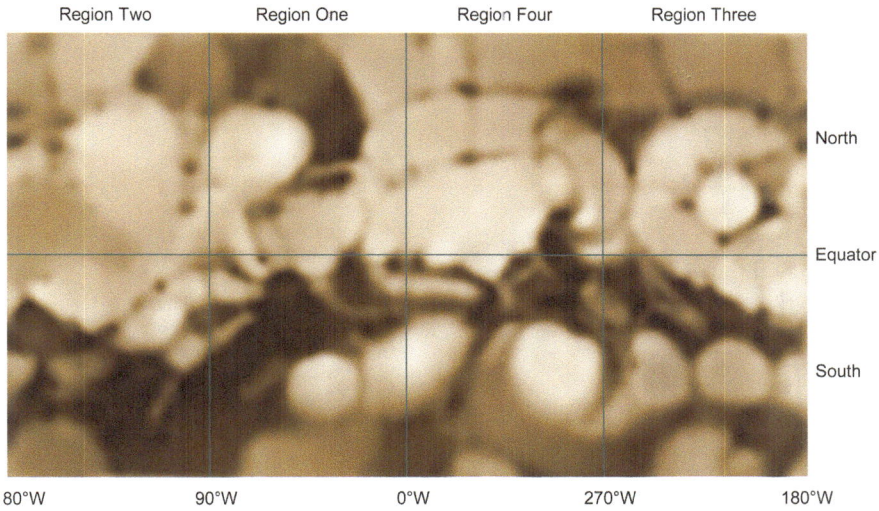

Albedo map of Mars (based on Antoniadi's map), north at top, showing the four regions described below (Credit: Grego)

8.1.1 Region One: 0–90°W

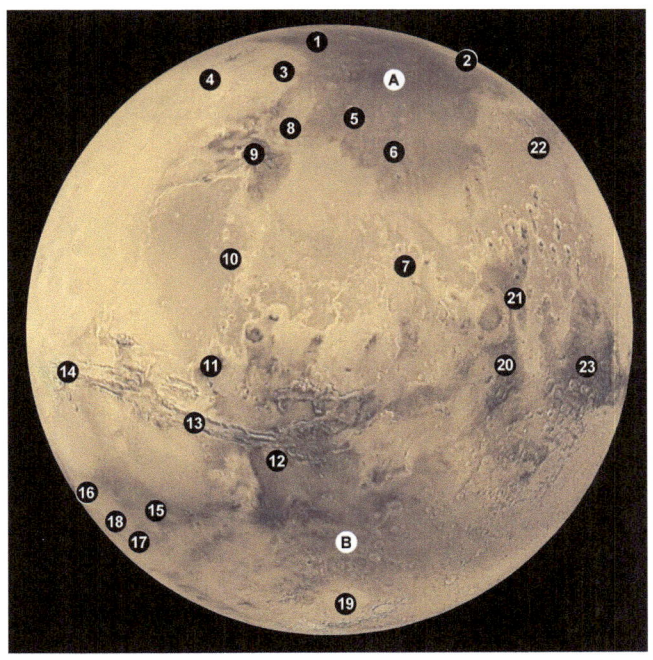

Hemisphere of Region One, centered on the equator at 45°W, north at top, with labeled albedo features mentioned in the text. Key (in order of first mention): *A*, Mare Acidalium; *1*, Baltia; *2*, Iaxartes; *3*, Tanais, Nix Tanaica; *4*, Tempe; *5*, Achillis Pons; *6*, Niliacus Lacus; *7*, Chryse; *8*, Nilokeras; *9*, Lunae Lacus; *10*, Ganges; *11*, Juventae Fons; *12*, Aurorae Sinus; *13*, Agathodaemon; *14*, Tithonius Lacus; *15*, Nectar; *16*, Solis Lacus; *17*, Bosporos; *18*, Thaumasia; *B*, Mare Erythraeum; *19*, Argyre; *20*, Margaritifer Sinus; *21*, Oxus; *22*, Siloe Fons; *23*, Sinus Meridiani. Letters refer to named maria (Credit: NASA/Google Earth/Grego)

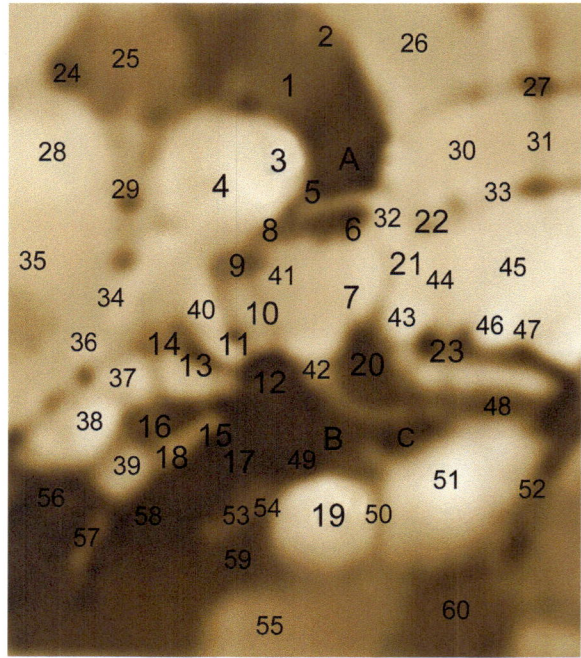

Albedo map of Region One centered on the equator at 45°W (70°N–70°S, 315–135°W), north at top. Additional labeled features (not mentioned in Region One text): *24*, Maeotis Palus; *25*, Nerigos; *26*, Ortygia; *27*, Arethusa Lacus; *28*, Arcadia; *29*, Ceraunius; *30*, Cydonia; *31*, Arnon; Oxia; *33*, Deuteronilus; *34*, Tharsis; *35*, Nix Olympica; *36*, Lux; *37*, Phoenicis Lacus; *38*, Claritas; *39*, Foelix Lacus; *40*, Candor; *41*, Xanthe; *42*, Eos; *43*, Aram; *44*, Thymiamata; *45*, Moab; *46*, Edom; *47*, Sigeus Portus; *48*, Pandorae Fretum; *C*, Mare Ionium; *49*, Depressio Erythraea; *50*, Argyroporos; *51*, Noachis; *52*, Hellespontus; *53*, Ogygis Regio; *54*, Nereidum Fretum; *55*. Argyre II; *56*, Aonius Sinus; *57*; Chrysokeras; *58*, Depressio Pontica; *59*, Campi Phlegraei; *60*, Depressiones Hellesponticae (Credit: Grego)

During perihelic oppositions, the broad dark region of **Mare Acidalium** (45°N, 30°W) in the north – one of the most clearly defined of Mars' marial areas -- always appears rather foreshortened near the northern limb. Its true shape and extent can be far more adequately appreciated during aphelic oppositions, when Mars' north pole is tilted towards the Sun. At such times, it is an easy feature to discern through a 100 mm telescope. At is largest, Mare Acidalium appears as a broad, well-defined dark mottled area, elongated north–south over around 25° of latitude from 60°N to 35°N. Its northwestern margin blends into the area of **Baltia** (60°N, 50°W) with a couple of dark wisps near **Iaxartes** (58°N, 20°W) from its far northern edge and **Tanais** (50°N, 70°W) extending from its western margin. Mare Acidalium's southern margin is usually rather more clearly defined, often appearing sharpest along its border with the bright circular area of **Tempe** (40°N, 70°W) to the northwest. Vague dusky streaks can sometimes be traced across Tempe, along with some brighter patches, including **Nix Tanaica** (52°N, 55°W) near the eastern edge of Tempe, a feature which can appear brilliant due to the formation of orographic cloud.

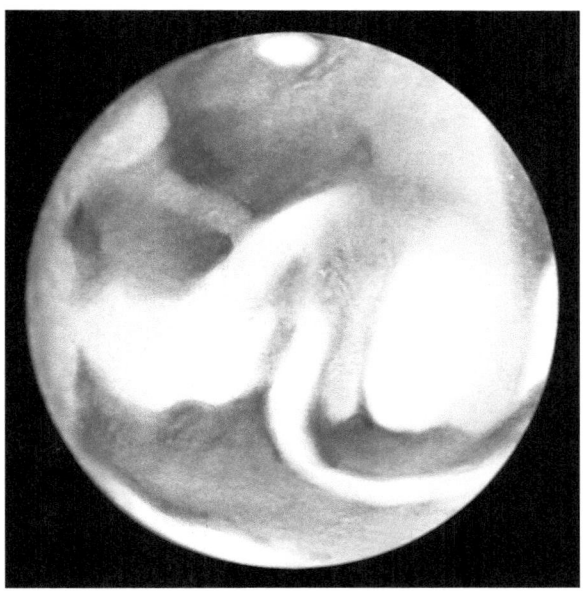

1999 May 6, 23 h UT. CM 12°. P 37°. Tilt 20.3°. Phase 99%. Diameter 16.1″. Magnitude −1.5. Margaritifer Sinus approaches the CM, a dusky streak from its northern tip extending to a hazy but easily seen Oxus and Siloe Fons. Mare Erythraeum is clearly defined on both its northern and southern borders, with a bright Argyre on the southern limb. Sinus Aurorae is visible, extending to Juventae Fons near the following limb. Sinus Meridiani and Sinus Sabaeus are dark, clearly separated from Margaritifer Sinus by a bright, well-defined Aram. Two dusky extensions proceed north of Sinus Meridiani from its two 'forks' across to a dusky Eden. Mare Acidalium is well-presented and mottled, with a bright Nix Cydonea on the CM, just north of the disk's center. Achillis Pons separates southern Mare Acidalium from a bold Niliacus Lacus, and Nilokeras lies near the following limb. Cydonia and Ortygia are fairly dark, and Dioscuria somewhat brighter on the preceding limb. A dark, bold Sinus Sabaeus extends to the preceding limb, and Edom and Sygeus Portus are visible. On the preceding limb, Ismenius Lacus is dark, bordered to its south by a bright Arabia. The north polar cap is small and brilliant. 150 mm Newtonian, 200×, integrated light (Credit: Grego)

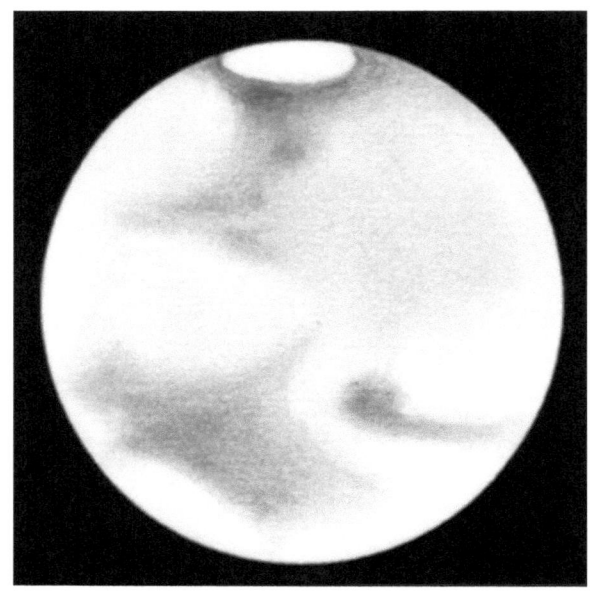

2010 February 1, 00 h UT. CM 21°. P 357°. Tilt 14.5°. Phase 99%. Diameter 14.1 ″. Magnitude −1.3. A well-defined Margaritifer Sinus is on the CM. Mare Erythraeum is darker towards its clear-cut southern border with a bright Argyre. Sinus Meridiani and Sinus Sabaeus are dark, clearly separated from Margaritifer Sinus by Aram. Faint shading lies north of Sinus Meridiani in the region of Eden. Mare Acidalium is well-presented, clearly defined at its northern edge by the north polar cap. Dusky spots within Mare Acidalium are visible, and a faint Achillis Pons separates it from Niliacus Lacus. Nilokeras is also visible. Tempe slightly brighter than Oxia and Eden. 300 mm Newtonian, 260×, integrated light (Credit: Paul Stephens)

Extending from eastern Tempe to define the southern edge of Mare Acidalium, a wide, light toned linear tract called **Achillis Pons** (38°N, 35°W) is frequently visible. Immediately to its south, the dark wedge of **Niliacus Lacus** (30°N, 30°W) reaches southwest for around 30°, separating southern Tempe from another bright region, **Chryse** (10°N, 30°W). A dark extension of Niliacus Lacus, **Nilokeras** (30°N, 55°W) is often observed to be double – it was once cited as the most obvious of the so-called Martian 'canals.' Nilokeras joins with a small dark patch, **Lunae Lacus** (20°N, 65°W). Lunae Lacus is rarely prominent, but several dusky swathes emanate from it, most notably **Ganges** (5°N, 58°W) which extends south to the Martian equator. Here it meets a small, well-defined dark spot, **Juventae Fons** (5°S, 63°W), usually visible through a 150 mm telescope at a favorable apparition. In good circumstances, the feature appears to be joined by an exceedingly narrow causeway to the delicate frog's hand-shaped **Aurorae Sinus** (15°S, 50°W) from which spreads a dusky web of tracts towards the north and west. Of these, **Agathodaemon** (10°S, 78°W) curves for nearly 15° of longitude across to **Tithonius Lacus** (5°S, 85°W) a dark spot usually somewhat larger and easier to discern than Juventae Fons. Agathodaemon and Tithonius Lacus actually mark the position of much of the magnificent Valles Marineris rift valley system.

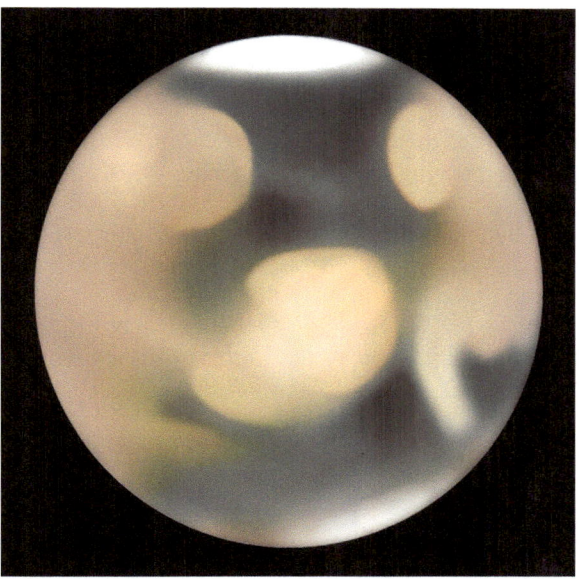

2010 January 30, 00:35 UT. CM 47°. P 357°. Tilt . Phase 100%. Diameter 14.1″. Magnitude −1.3. Mars at opposition. The north polar cap is brilliant. Just past the CM, Mare Acidalium is dark, prominent and well-defined. Nilokeras is easily visible, and spreads towards Lunae Lacus. Chryse is central to the disk and fairly bright. Aurorae Sinus is dusky but ill-defined. Sinus Margaritifer is dark, extending north to Niliacus Lacus and clearly separated by a bright Aram from a prominent Sinus Meridiani near the eastern limb. Agathodaemon and Melas Lacus are clearly seen, and Argyre at the southern limb is bright. 300 mm Newtonian, 200×, integrated light and yellow (W12) filter (Credit: Grego)

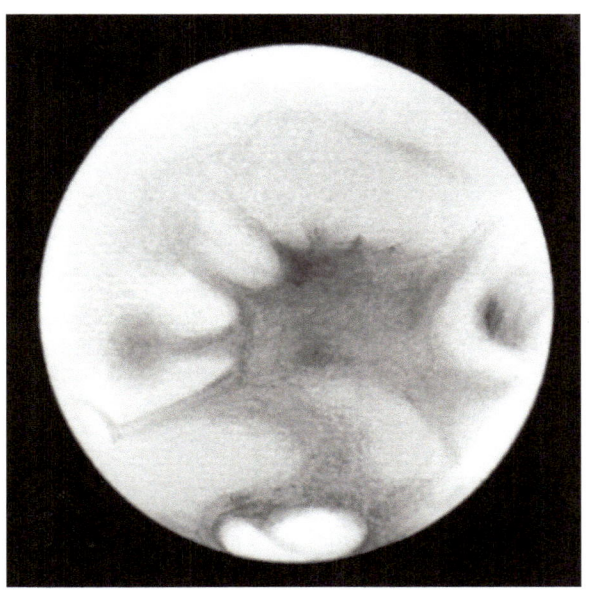

2003 August 20, 00:45 UT. CM 51°. P 345°. Tilt −19.0°. Phase 99%. Diameter 24.8″. Magnitude −2.7. A dark Aurorae Sinus lies on the CM. Extending from it is a dark Juventae Fons and Agathodaemon leading to an indistinct Melas Lacus. Nectar extends west to join Solis Lacus, which is nicely defined by Thaumasia and Foelix Lacus. Ogygis Regio and Campi Phlegraei extend south, bordering Argyre I and II. The south polar cap is brilliant and darkly bordered; note the indentation of the cap and the narrow lane extending to the pole – this is Rima Australe, which appears as the southern polar cap retreats in the Martian summer. Sinus Meridiani, bordered by Aram, nears the preceding limb. Note also traces of a foreshortened Niliacus Lacus north of a dull Chryse. 300 mm Newtonian, 260×, integrated light (Credit: Paul Stephens)

Southwest of Aurorae Sinus an occasionally broad dark tract called **Nectar** (28°S, 72°W) runs west to join one of Mars' most prominent features, **Solis Lacus** (28°S, 90°W), a roughly circular patch which varies in extent and intensity from one apparition to the next. At its largest, Solis Lacus covers more than 20° in longitude. It sits just south of the central position within a brighter area bounded in the south by the dark curving tract of **Bosporos** (34°S, 64°W) and usually separated from it by the light-toned **Thaumasia** (35°S, 85°W). Sometimes the southern edge of Solis Lacus blends into a darker than usual Thaumasia, making its outline difficult to trace, but it is always pretty clearly defined in the north. Occasionally Solis Lacus appears as a mottled collection of dark, ill-defined patches, but sometimes it can appear as a striking single spot (especially in poorer than average seeing conditions or through a small telescope) justifying its unofficial name of the 'Eye of Mars'.

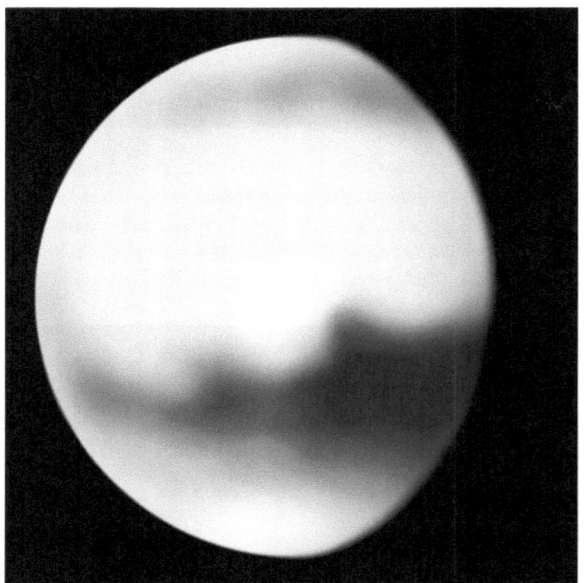

2007 September 16, 02:00 UT. CM 88°. P 331°. Tilt 0.7°. Phase 86%. Diameter 8.8 ″. Magnitude 0.1. In this observation early in the 2007–2008 apparition when Mars was very small and showing a large phase, Solis Lacus is near the center of the disk, but not well seen. Aurorae Sinus is broad and dark near the preceding limb, with Agathodaemon visible. A bright area is visible northeast of Solis Lacus. Mare Sirenum less distinct towards following limb. Bright south and north polar areas, no distinct caps seen. Slight shading in Baltia and Nerigo is observed. 150 mm Newtonian, 235×, integrated light (Credit: Grego)

South of Aurorae Sinus, stretching a quarter of the way around Mars from around 0°W to 90°W, between latitudes 20°S and 40°, is **Mare Erythraeum** (25°S, 40°W) which appears as a collection of dark patches. In the south, Mare Erythraeum fades and embraces the large and usually light toned, clearly-defined circular patch of **Argyre** (45°S, 25°W) a major ancient impact basin.

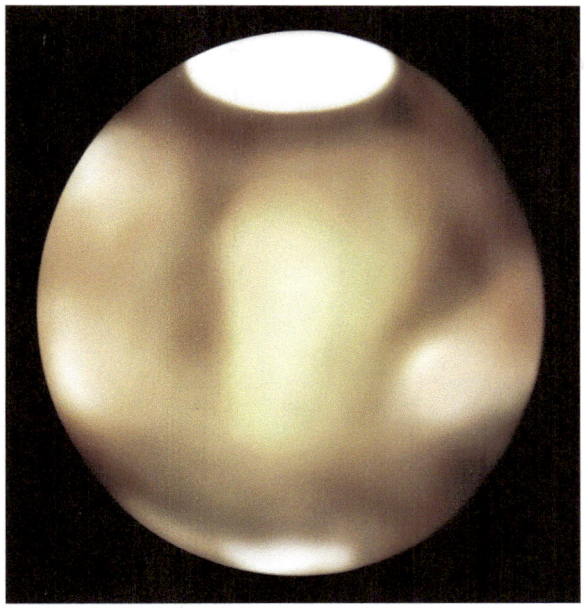

2009 December 19, 03:15 UT. CM 99°. P 3°. Tilt 19.0°. Phase 94%. Diameter 11.5″. Magnitude −0.5. A bright Tharsis, central to the disk, is bordered to the south by a very indistinct Solis Lacus. Towards the preceding limb are Lunae Lacus and Nilokeras, with Mare Acidalium almost about to enter the evening terminator. Western parts of Chryse, to its south, is bright, underscored by a dusky Mare Erythraeum, with Mare Sirenum on the following limb. There's brightening of the southern limb, along with a bright Amazonis and Arcadia on the following limb. Approaching the CM, Ceraunius is clearly visible, extending north to Nerigo and the dark border of a large north polar cap. 200 mm SCT, 285×, integrated light (Credit: Grego)

Mare Erythraeum darkens in the north at around 20°W, and links with **Margaritifer Sinus** (10°S, 25°W), a dark narrow V-shaped feature which extends and narrows north of the equator, curving slightly eastward around the margin of Chryse. Margaritifer Sinus usually shows a paler tone than Aurorae Sinus to its west, but its northern tip sometimes appears very pointed and extends along a tract called **Oxus** (20°N, 12°W) to the dusky node of **Siloe Fons** (33°N, 8°W) southeast of Mare Acidalium. East of Margaritifer Sinus, on the Martian prime meridian, lies the prominent dark feature **Sinus Meridiani** (5°S, 0°W), also discussed in Region Four below.

8.1.2 Region Two: 90–180°W

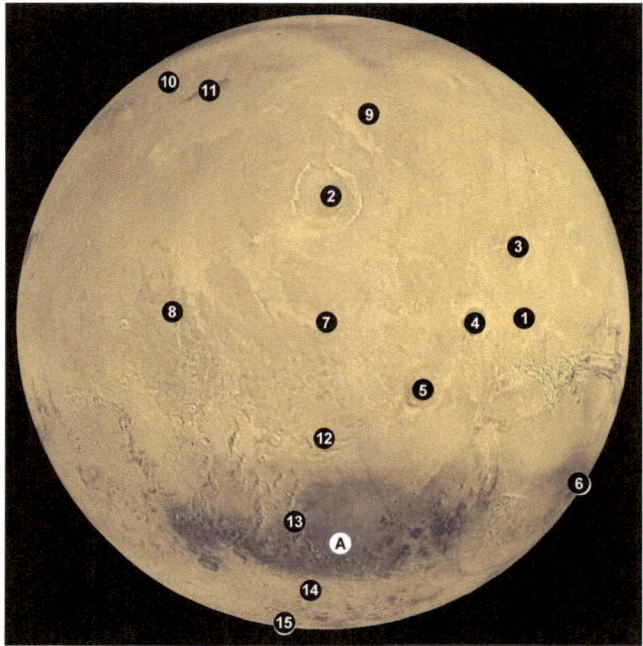

Hemisphere of Region Two, centered on the equator at 135°W, north at top, with labeled albedo features mentioned in the text. Key (in order of first mention): *1*, Tharsis; *2*, Nix Olympica; *3*, Ascraeus Lacus; *4*, Pavonis Lacus; *5*, Arsia Silva; *6*, Solis Lacus; *7*, Nodus Gordii; *8*, Amazonis; *9*, Arcadia; *10*, Castorius Lacus; *11*, Propontis; *A*, Mare Sirenum; *12*, Memnonia; *13*, Fusca Depressio; *14*, Phaethontis; *15*, Thyle I. Letters refer to named maria (Credit: NASA/Google Earth/Grego)

Albedo map of Region Two centered on the equator at 135°W (70°N–70°S, 45–225°W), north at top. Additional labeled features (not mentioned in text): *16*, Panchaia; *17*, Scandia; *18*, Maeotis Palus; *19*, Nerigos; *20*, Baltia; *21*, Stymphalius Lacus; *22*, Cebrenia; *23*, Elysium; *24*, Styx; *25*, Trivium Charontis; *26*, Cerberus; *27*, Aeolis; *28*, Zephyria; *29*, Orcus; *30*, Phlegra; *31*, Euxinus Lacus; *32*, Ceraunius; *33*, Tempe; *34*, Nix Tanaica; *35*, Nilokeras; *36*, Lunae Lacus; *37*, Candor; *38*, Ganges; *39*, Juventae Fons; *40*, Tithonius Lacus; *41*, Melas Lacus; *42*, Lux; *43*, Phoenicis Lacus; *44*, Claritas; *B*, Mare Cimmerium; *45*, Eridania; *46*, Electris; *C*, Mare Chronium; *47*, Siomis; *48*, Palinuri Fretum; *49*, Thyle II; *50*, Aonius Sinus; *51*, Depressiones Aoniae; *52*, Chrysokeras; *53*, Dia; *54*, Foelix Lacus; *55*, Thaumasia; *56*, Nectar; *57*, Aurorae Sinus; *58*, Ogygis Regio; *59*, Campi Phlegraei; *60*, Argyre II (Credit: Grego)

Tharsis (0°N, 100°W), home to the largest volcanic province on Mars, straddles the equator between around 80°W and 120°W. Despite the gigantic nature of these volcanoes, it is difficult to see them directly through backyard telescopes. **Nix Olympica** (Olympus Mons, 21°N, 127°W) is often visible because of the bright orographic clouds that form near its summit – long before its mountainous nature was known, the feature merited its highly appropriate name -- truly an Olympic mountain, fit for the home of the gods! Under excellent circumstances, three faint dusky spots can be seen near Nix Olympica – **Ascraeus Lacus** (12°N, 105°W), **Pavonis Lacus** (2°N, 113°W) and **Arsia Silva** (8°S, 121°W) – large sister volcanoes forming a northeast-southwest line immediately between Nix Olympica and Solis Lacus. Bright orographic clouds, caused when air laden with water vapor is pushed to high altitudes, occur over all of these features and are sometimes easily seen through quite small telescopes without the use of filters. A W-shaped cloud formation, sometimes seen over the Tharsis volcanoes in late southern spring afternoons, can appear remarkably bright without filters; similar clouds can occur at around the same time over Elysium (see Region Three).

2010 February 23, 21:40 UT. CM 144°. P 352°. Tilt 12.5°. Phase 97%. Diameter 12.6″. Magnitude −0.8. A subtly shaded Amazonis is on the center of the disk. In the south, Mare Sirenum is fairly dusky, followed by Mare Cimmerium towards terminator. There is considerable brightening towards the preceding limb in the Tharsis region. The north polar cap is extensive and bordered by a dusky collar, widening towards the following limb. 200 mm SCT, 250×, integrated light (Credit: Grego)

A host of other subtle and often ill-defined features may be glimpsed in the region, including the extensive patch of **Nodus Gordii** (0°N, 135°W) and various ill-defined dusky zones within **Amazonis** (0°N, 140°W) and **Arcadia** (45°N, 100°W), both light-toned zones north of the equator. The far northern latitudes are considerably less intensely shaded than in the southern hemisphere, but to the west of Arcadia there is a patchwork of smaller dusky features, including **Castorius Lacus** (52°N, 155°W) and **Propontis** (45°N, 185°W).

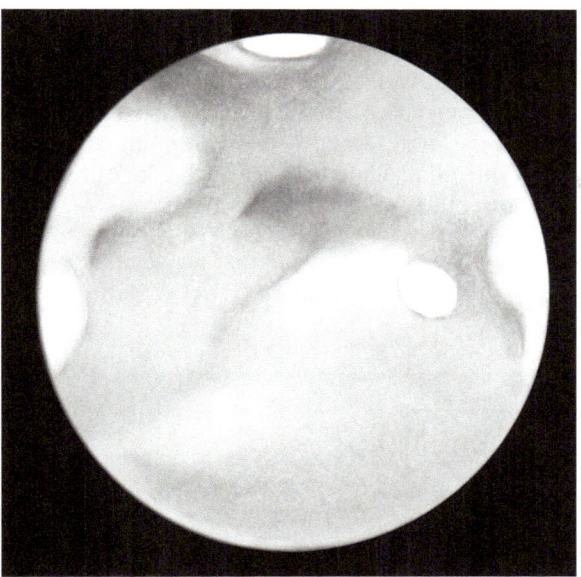

1997 March 31, 21:30 UT. CM 168°. P 23°. Tilt 23.4°. Phase 99%. Diameter 13.9″. Magnitude −1.1. Amazonis, central to the disk, appears dusky, the darkest area being Propontis. To its west, near the following limb, is a bright, well-defined Elysium, with an easily visible Trivium Charontis and Cerberus, with Aethiopis bright and well-defined on the limb. Further north, Panchaia is dark and extends to the following limb. A dusky tract runs south from Propontis, across Zephyria, towards a poorly distinguished Mare Cimmerium near the southern limb; Mare Sirenum is bland and virtually indiscernible. Nix Olympica is a distinct bright spot, and western Tharsis is crossed by a dusky tract, but at the preceding limb another bright area is visible over the Tharsis volcanoes. The main body of the north polar cap is brilliant, separated from a less bright Olympia by Rima Australis. 225 mm Newtonian, 250×, integrated light (Credit: Grego)

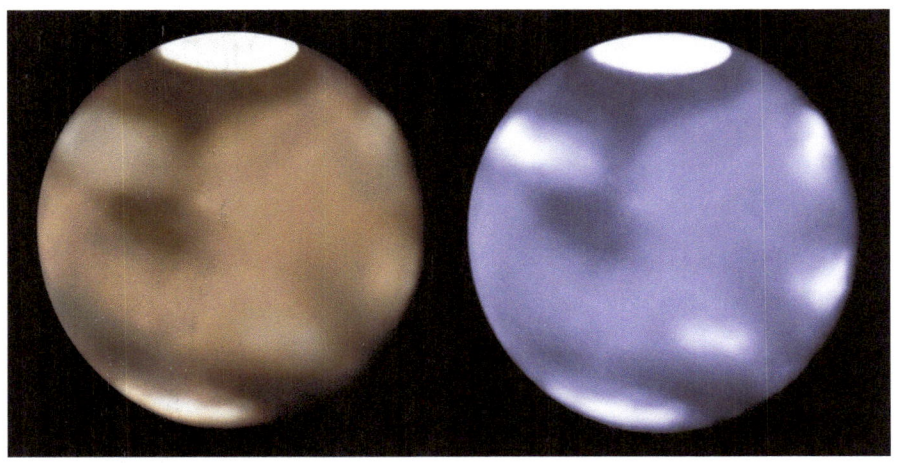

2010 January 13, 23:30 UT. CM 171°. P 1°. Tilt 16.8°. Phase 99%. Diameter 13.7″. Magnitude −1.1. In both integrated light and in *yellow* and *blue* filters, the large north polar cap is brilliant with a dusky border extending south through Phlegra to a dusky but poorly defined Trivium Charontis. In *blue* light several other bright areas stand out prominently — one in the Elysium area to the northwest of Trivium Charontis, another in the Memnonia region bordering Mare Sirenum, one in the southern Tharsis region near the terminator and another near the terminator but further north in the vicinity of Ascraeus Lacus. The Electris region along the southern limb was also bright. 200 mm SCT, 250×, integrated light and yellow W12 (left) and blue W80A (right) (Credit: Grego)

South of the equator, the dark crescent of **Mare Sirenum** (30°S, 155°W) borders the brighter region of **Memnonia** (20°S, 150°W); at those times when Mare Sirenum is on the central meridian, it is likely to be the darkest feature visible, with **Fusca Depressio** (33°S, 147°W), a small lobe on its northeastern border, usually the most intensely dark part of the mare. South of Mare Sirenum are a number of broad, light toned regions, including **Phaethontis** (50°S, 155°W), while further south, **Thyle** (70°S, 180°W) extends towards the south polar cap.

8.1.3 Region Three: 180–270°W

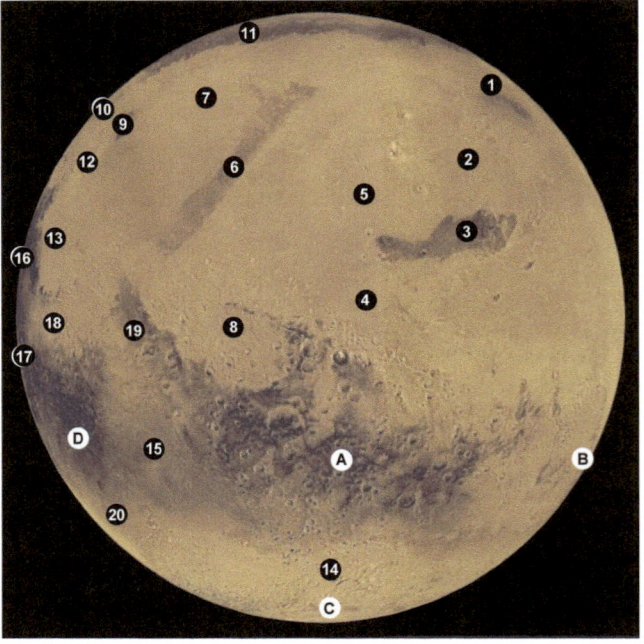

Hemisphere of Region Three, centered on the equator at 225°W, north at top, with labeled albedo features mentioned in the text. Key (in order of first mention): *1*, Propontis; *2*, Phlegra; *3*, Trivium Charontis; *4*, Cerberus; *5*, Elysium; *6*, Hyblaeus; *7*, Aetheria; *8*, Aethiopis; *9*, Nodus Alcyonius; *10*, Casius; *11*, Utopia; *12*, Thoth-Nepenthes; *13*, Moeris Lacus; *A*, Mare Cimmerium; *B*, Mare Sirenum; *14*, Eridania; *C*, Mare Chronium; *15*, Hesperia; *D*, Mare Tyrrhenum; *16*, Syrtis Major; *17*, Crocea; *18*, Libya; *19*, Syrtis Minor; *20*, Ausonia Borealis; *21*, Ausonia Australis. Letters refer to named maria (Credit: NASA/Google Earth/Grego)

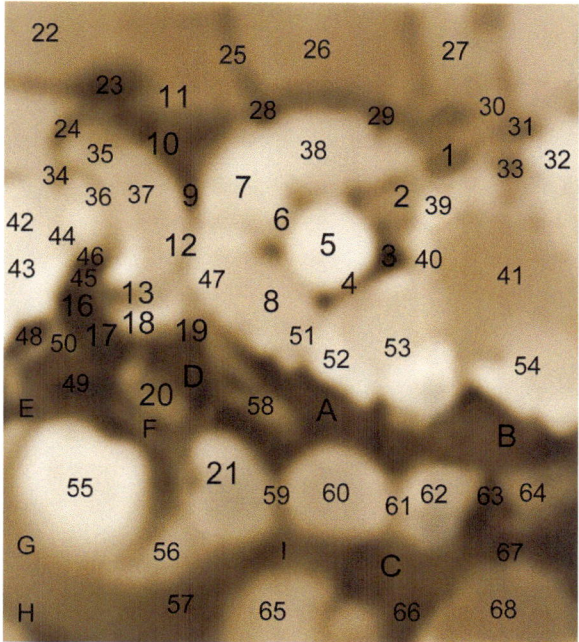

Albedo map of Region Three centered on the equator at 225°W (70°N–70°S, 135–315°W), north at *top*. Additional labeled features (not mentioned in text): *22*, Cecropia; *23*, Copais Palus; *24*, Boreosyrtis; *25*, Cydnus; *26*, Panchaia; *27*, Scandia; *28*, Stymphalius Lacus; *29*, Hecates Lacus; *30*, Diacria; *31*, Castorius Lacus; *32*, Arcadia; *33*, Euxinus Lacus; *34*, Boreus Pons; *35*, Umbra; *36*, Nilosyrtis; *37*, Neith Regio; *38*, Cebrenia; *39*, Azania; *40*, Orcus; *41*, Amazonis; *42*, Arabia; *43*, Aeria; *44*, Astusapes; *45*, Arena; *46*, Nili Sinus; *47*, Amenthes; *48*, Deltoton Sinus; *49*, Iapygia; *50*, Oenotria; *51*, Cyclopia; *52*, Aeolis; *53*, Zephyria; *54*, Memnonia; E, Mare Ionium; F, Mare Hadriacum; G, Mare Amphitrites; H, Mare Australe; *55*, Hellas; *56*, Chersonesus; *57*, Promethei Sinus; *58*, Hesperia; *59*, Xanthes; *60*, Eridania; *61*, Scamander; *62*, Electris; *63*, Simois; *64*, Phaethontis; I, Mare Chronium; *65*, Thyle II; *66*, Ulyxis Fretum; *67*, Palinuri Fretum; *68*, Thyle I (Credit: Grego)

Extending from Propontis, spanning 20° of latitude, the ill-defined tract of **Phlegra** (31°N, 188°W) meets the dusky **Trivium Charontis** (20°N, 198°W). **Cerberus** (15°N, 205°W), often seen as a prominent wide dark tract, extends southwest of Trivium Charontis and skirts the southern edge of the bright circular **Elysium** (25°N, 210°W) a volcanic plateau which is often host to bright orographic cloud formations. These features are often easily visible through a 100 mm telescope. **Hyblaeus** (35°N, 235°W) marks the western edge of Elysium; in this chart it appears prominent, dark and linear, but in most apparitions it is usually far less well defined. Further west, faint markings can often be seen in the regions of **Aetheria** (40°N, 230°W) and **Aethiopis** (10°N, 230°W). Nearby, **Nodus Alcyonius** (35°N, 257°W) forms a sometimes clearly defined dusky node, into which spiral several ill-defined tracts, such as **Casius** (40°N, 260°W) emanating south from **Utopia** (50°N, 250°W) and **Thoth-Nepenthes** (20°N, 260°W) linking **Moeris Lacus** (8°N, 270°W) with Nodus Alcyonius.

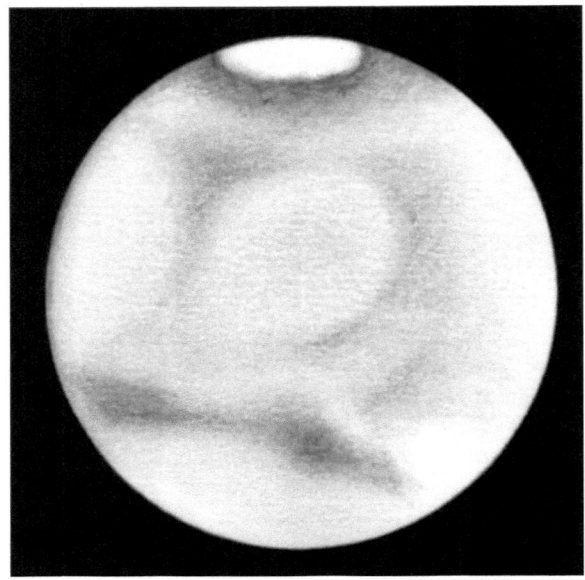

2010 February 16, 22:45 UT. CM 223°. P 353°. Tilt 12.9°. Phase 98%. Diameter 13.2. Magnitude −0.9. A dull but clearly defined Elysium is almost central on the disk. To its west can be seen Nodus Alcyonius and Casius, with Propontis to its east and streaks in the Cerberus-Trivium Charontis area. A darker Mare Tyrrhenum and Mare Cimmerium lies towards the southern limb. The north polar cap is bright and has a dark border. 300 mm Newtonian, 260×, integrated light (Credit: Paul Stephens)

Just south of the equator, Cerberus joins with the northwestern reaches of **Mare Cimmerium** (20°S, 220°W) a broad dusky arc stretching almost 70° in longitude from Mare Sirenum to 245°W. Mare Cimmerium is an easy object through small telescopes. It is darkest and most clearly defined along its northern margin, and displays considerable detail through a 200 mm telescope under good seeing conditions. **Eridania** (45°S, 220°W) is occasionally easy to discern as a light patch south of Mare Cimmerium, and to its south **Mare Chronium** (58°S, 210°W) is marked by a hazy streak.

2010 January 4, 22:30 UT. CM 236°. P 2.6°. Tilt 17.9°. Phase 97%. Diameter 13.0 ". Magnitude −0.9. Integrated light (left) and yellow W12 filter. The north polar cap is brilliant with a dark border. Syrtis Major is near the western limb, with brightening at the limb along western edge of Syrtis Major and along the southern limb. There is darkening and a pronounced V-shape to Casius. Mare Tyrrhenum, Hesperia and Mare Cimmerium are all clearly visible. Dusky, ill-defined patches seen in Elysium. In the yellow filter brightening to the east and west of Syrtis Major is noticeable, so too is a brightening to the east of Casius and along the planet's southern limb. A vague dusky extension of Elysium to the west towards Casius is also discernable in the yellow filter. 200 mm SCT, 285× (Credit: Grego)

Hesperia (20°S, 240°W) a light zone, separates Mare Cimmerium from **Mare Tyrrhenum** (20°S, 255°W). Sometimes the division between the two maria is not too distinct, but Mare Tyrrhenum is usually the darker of the two, although it can display considerable patchiness. Western Mare Tyrrhenum joins with northern **Syrtis Major** (10°N, 290°W) at an occasionally well-defined junction called **Crocea** (5°S, 288°W); bounded by **Libya** (0°N, 270°W), Mare Tyrrhenum's northern edge extends into a narrow pointed feature called **Syrtis Minor** (8°S, 260°W) whose northern tip just touches the equator. **Ausonia Borealis** (23°S, 275°W), a region south of Mare Tyrrhenum, sometimes appears rather bright, but at other times is indistinct; this area extends south to **Ausonia Australis** (40°S, 250°W).

8.1.4 Region Four: 270–360°W

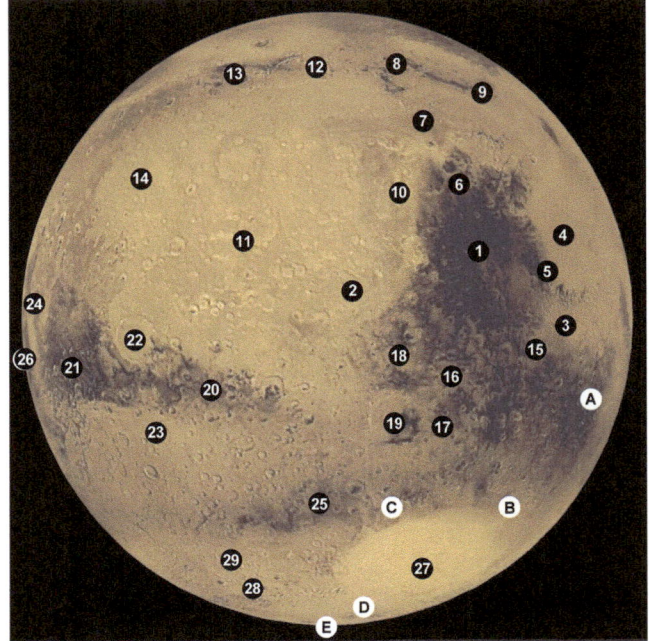

Hemisphere of Region Four, centered at 315°W, north at top, with labeled albedo features mentioned in the text. Key (in order of first mention): A, Mare Tyrrhenum; *1*, Syrtis Major; *2*, Aeria; *3*, Libya; *4*, Isidis Regio; *5*, Moeris Lacus; *6*, Arena; *7*, Nilosyrtis; *8*, Boreosyrtis; *9*, Casius; *10*, Astusapes; *11*, Arabia; *12*, Protonilus; *13*, Ismenius Lacus; *14*, Eden; *15*, Crocea; *16*, Oenotria; *17*, Iapygia; *18*, Deltoton Sinus; *19*, Huygens; *20*, Sinus Sabaeus; *21*, Sinus Meridiani; *22*, Sigeus Portus, Schiaparelli; *23*, Deucalionis Regio; *24*, Thymiamata; *25*, Pandorae Fretum; *26*, Margaritifer Sinus; *27*, Hellas; B, Mare Hadriacum; C, Mare Ionium; D, Mare Amphitrites; E, Mare Australe; *28*, Hellespontus; *29*, Noachis. Letters refer to named maria (Credit: NASA/Google Earth/Grego)

Albedo map of Region Four centered on the equator at 315°W (70°N–70°S, 225–45°W), north at top. Additional labeled features (not mentioned in text): *F*, Mare Acidalium; *30*, Achillis Pons; *31*, Niliacus Lacus; *32*, Nix Cydonea; *33*, Cedron; *34*, Ortygia; *35*, Kison; *36*, Cecropia; *37*, Copais; *38*, Utopia; *39*, Cydnus; *40*, Sithonius Lacus; *41*, Cydonia; *42*, Arethusa Lacus; *43*, Arnon; *44*, Dioscuria; *45*, Boreus Pons; *46*, Asclepii Pons; *47*, Neith Regio; *48*, Nodus Alcyonius; *49*, Aetheria; *50*, Thoth-Nepenthes; *51*, Morpheos Lacus; *52*, Hyblaeus; *53*, Hephaestus; *54*, Aethiopis; *55*, Cyclopia; *56*, Oxia; *57*, Chryse; *58*, Moab; *59*, Edom; *60*, Syrtis Minor; *61*, Hesperia; *G*, Mare Cimmerium; *H*, Mare Chronium; *I*, Mare Erythraeum; *62*, Pyrrhae Regio; *63*, Depressio Erythraea; *64*, Vulcani Pelagus; *65*, Argyre I; *66*, Argyroporos; *J*, Mare Oceanidum; *67*, Depressiones Hellesponticae; *68*, Centauri Lacus; *69*, Ausonia Borealis; *70*, Ausonia Australis; *71*, Ledae Pons; *72*, Malea Promontorium; *73*, Chersonesus; *74*, Promethei Sinus (Credit: Grego)

Mare Tyrrhenum flows northwest towards the equator, where it blends with the southeastern corner of **Syrtis Major**, a very prominent broad dark wedge visible in the smallest telescopes at opposition. Syrtis Major protrudes northwards across the equator to around 20°N latitude. Its western edge, bordering the bright area of **Aeria** (10°N, 310°W), is most often the darkest and most clearly defined part of the feature; its eastern edge, bordering **Libya** and **Isidis Regio** (20°N, 275°W), displays marked seasonal variations. A clearly defined bright patch within Isidis Regio, just north of the protruding Moeris Lacus, is sometimes striking, even through small telescopes. Syrtis Major often appears distinctly mottled, and its northern section occasionally appears detached from the rest by a narrow bright lane called **Arena** (13°N, 293°W). During some apparitions Syrtis Major's northern end appears sharp and pointed, but at other apparitions it appears somewhat lobate, blunted or angular. Syrtis Major's northern tip sometimes appears to curve and extends further northwards along a prominent curved streak called **Nilosyrtis** (42°N, 290°W), where it meets with the dusky tract of **Boreosyrtis** (55°N, 290°W) which links with Casius further east. At some apparitions, the northwestern edge of Syrtis Major extends to a dark point, from which a fine dark streak called **Astusapes** (25°N, 298°W) is sometimes visible.

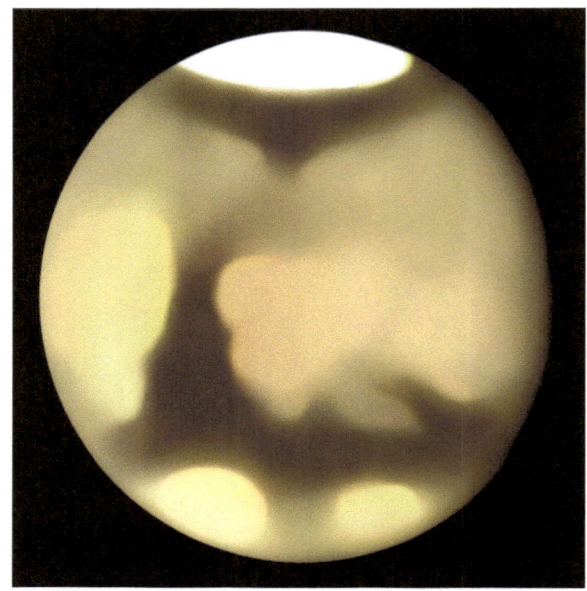

2010 January 7, 01:45 UT. CM 266°. P 2°. Tilt 17.7°. Phase 98%. Diameter 13.2″. Magnitude −0.9. Isidis Regio is near the center of the disk, followed by a prominent dark Syrtis Major. Faint indications of Astusapes and Nili Lacus are seen, along with slight projections to the east and west of Syrtis Major. Hellas is well-defined on the limb and is the brightest feature except the very sizeable north polar cap. Syrtis Minor, Mare Tyrrhenum and Mare Sirenum easily seen, with some brightening in Ausonia near the southern limb. Extremely faint but indistinct traces of Cerberus and Trivium Charontis. Utopia and Nodus Alcyonius are easily visible. 200 mm SCT, 200×, integrated light and yellow W12 filter (Credit: Grego)

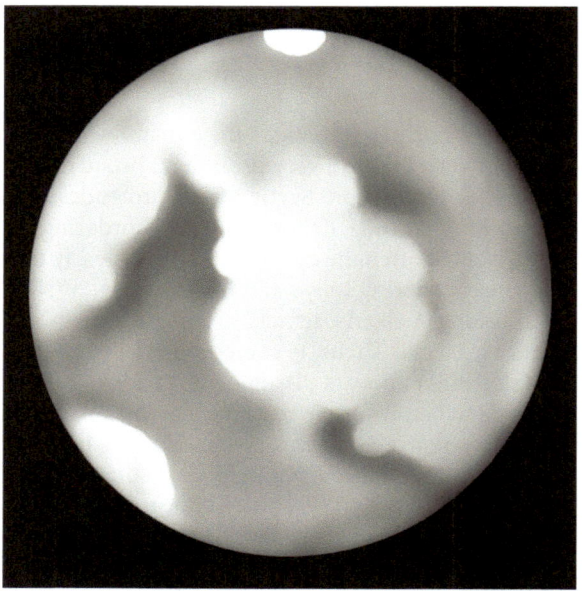

1999 April 19, 00:30 UT. CM 276°. P 38°. Tilt 16.6°. Phase 99%. Diameter 15.1″. Magnitude −1.4. A bright Libya region is near the center of the disk, followed by a prominent dark Syrtis Major. Astusapes is clearly visible, and so are projections to the east and west of Syrtis Major. Hellas is bright and well-defined on the limb. Syrtis Minor, Mare Tyrrhenum and Mare Sirenum easily seen, with some brightening in Ausonia near the southern limb. Cerberus is faint but traceable east of Elysium to a dusky, but poorly-defined Trivium Charontis. Zephyria bright on preceding limb. Nodus Alcyonius and Thoth-Nepenthes are visible, along with Boreus Pons near the following limb. The north polar cap is small but brilliant. 150 mm Newtonian, 200×, integrated light (Credit: Grego)

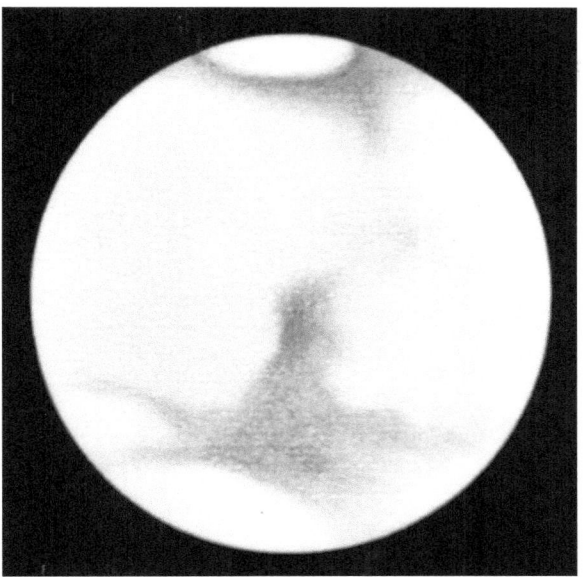

2010 February 9, 23:30 UT. CM 295°. P 355°. Tilt 13.5°. Phase 99%. Diameter Magnitude −1.1. Syrtis Major, almost central on the disk, is dark and well-defined along its western border. To its sides, Aeria and Isidis Regio are fairly muted. Moeris Lacus can be discerned east of Syrtis Major. Mare Tyrrhenum extends towards the preceding limb, while Sinus Sabaeus, Deucalionis Regio and Pandorae Fretum stretch towards the following limb. Hellas is quite bright and clearly defined by Pandorae Fretum. The north polar cap is brilliant, darkly bordered and shows a dusky extension towards the east, probably Nodus Alcyonius. 300 mm Newtonian, 260×, integrated light (Credit: Paul Stephens)

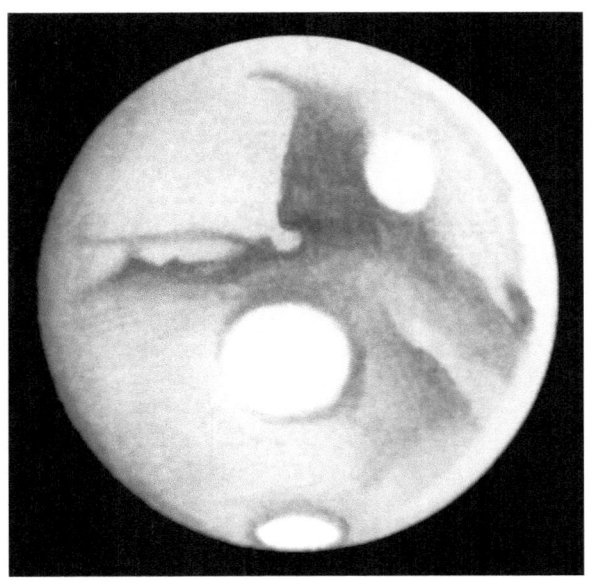

1988 October 30, 00:05 UT. CM 304°. P 333°. Tilt −24.3°. Phase 95%. Diameter 18.8″. Magnitude −1.9. Syrtis Major, dark and clearly defined, is leaving the CM. The delicate wisp of Astusapes is visible emanating from the northern tip of Syrtis Major. Libya is bright and Syrtis Minor is visible, but Moeris Lacus and Isidis Regio are not clearly defined. A darker region around Iapygia has a small semicircular bay, possibly the crater Huygens. Sinus Sabaeus is clearly delineated along its northern edge, and Sigeus Portus, the site of crater Schiaparelli, is evident; the terminator darkens towards the west, but Sinus Meridiani is not visible. A faint linear marking running north of and parallel to Sinus Sabaeus can be seen. Hellas, below center of disk, is bright and well-defined. Towards the brighter preceding limb can be seen a dusky Amenthes and dark protruding Cerberus III. The south polar cap is brilliant with a dark border, no irregularities observed along its edge. 300 mm Newtonian, 200×, integrated light (Credit: Grego)

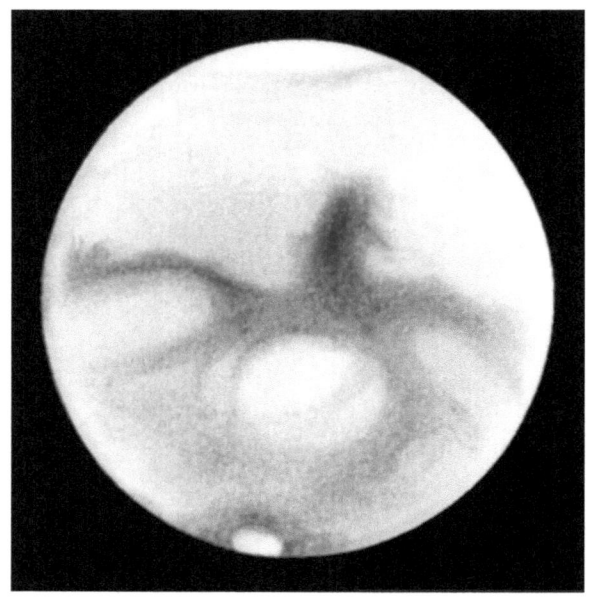

2005 November 30, 00:15 UT. CM 311°. P 321°. Tilt −18.9°. Phase 97%. Diameter 17.1 ″. Magnitude −1.7. A prominent Syrtis Major has moved off the CM. Moeris Lacus extends east of Syrtis Major. Mare Tyrrhenum extends towards the preceding limb, while Sinus Sabaeus, Deucalionis Regio and Pandorae Fretum stretch towards the following limb; The dual 'forked' appearance of Sinus Meridiani, near the following limb, is evident. Hellas, below center of disk, is bright and clearly defined by Pandorae Fretum in the north. Hellespontus in the west and Mare Hadriacum in the northeast. A dull wisp, Alpheus, crosses southern Hellas. Mare Australe surrounds a small but brilliant south polar cap. 300 mm Newtonian, 300×, integrated light (Credit: Paul Stephens)

Arabia (20°N, 330°W), a wide light-toned desert area, blends into the usually brighter region of Aeria. Its northern edge, at around 40°N, is defined by the dark tracts of **Protonilus** (42°N, 315°W) and **Ismenius Lacus** (40°N, 330°W). To their southwest, the desert region of **Eden** (30°N, 350°W) spans more than 40° latitude, but displays only very subtle tonal variations.

Syrtis Major's southern edge is often defined by Crocea at the border of Mare Tyrrhenum and **Oenotria** (5°S, 295°W), a light toned strip which borders the dusky patches of **Iapygia** (20°S, 295°W) and **Deltoton Sinus** (4°S, 305°W). A distinct semicircular notch in Iapygia at 14°S, 304°W, is sometimes easily visible through a 100 mm telescope. This bay of eastern Aeria actually marks the position of Huygens, an impact crater 467 km across. South of Aeria and extending west of Iapygia for some 40° of longitude is the prominent dark tract of **Sinus Sabaeus** (8°S, 340°W). More often at its darkest and most clearly delineated along its northern edge, this well-known feature joins with **Sinus Meridiani** (5°S, 0°W) the famous 'forked bay' whose dual prongs extend northwards across the equator, straddling the 0° line of longitude. A dark peninsula called **Sigeus Portus** (5°S, 335°W) is often visible midway along the northern edge of Sinus Sabaeus; a clearly discernable bay in this feature marks the position of another large Martian impact crater, the 459 km diameter **Schiaparelli**. Adjoining the southern edge of Sinus Sabaeus is the variable light-toned tract of **Deucalionis Regio** (15°S, 340°W) which flows around the western margin of Sinus Sabaeus into the bright desert region of **Thymiamata** (10°N, 10°W). Another dusky tract, **Pandorae Fretum** (25°S, 316°W) lies along the southern edge of Deucalionis Regio, and curves northwards into the dark V-shaped **Margaritifer Sinus** (10°S, 25°W).

Often the most clearly defined bright area on Mars, **Hellas** (40°S, 290°W) lies around 40° due south of Syrtis Major and is usually easily identifiable through a small telescope. Its margin is bordered by the dusky **Mare Hadriacum** (40°S, 270°W) in the northeast, **Mare Ionium** (25°S, 310°W) in the northwest, **Mare Amphitrites** (55°S, 310°W) in the southwest and **Mare Austral** (60°S) in the south. A dark tract called **Hellespontus** (50°S, 325°W) curves southwest from Mare Ionium, defining the southern border of the light-toned desert region of **Noachis** (45°S, 330°W).

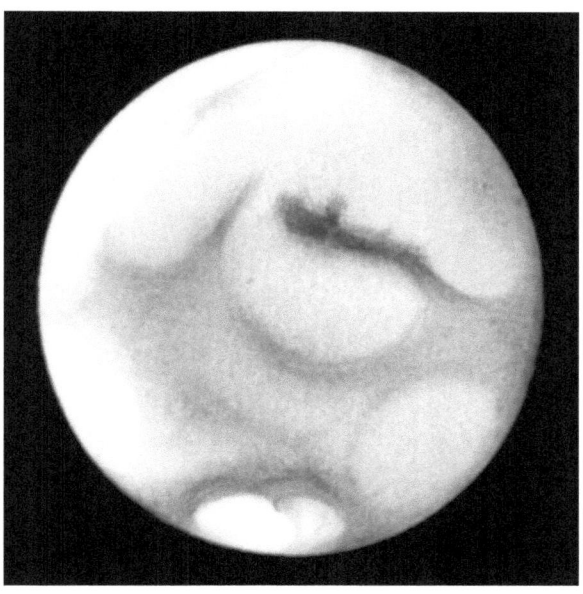

2003 July 20, 02:15 UT. CM 353°. P 344°. Tilt −20.5°. Phase 93%. Diameter 20.1˝. Magnitude −2.0. A prominent dual-forked Meridiani Sinus is near the CM. Sinus Sabaeus is also prominent and shows a hint of Sigeus Portus (location of the crater Schiaparelli). Deucalionis Regio is broad, and Margaritifer Sinus extends north in a long tapering point; to its west, near the following limb, is a hint of Aurorae Sinus. Noachis, approaching the preceding limb, is clearly outlined but is about the same tone as Deucalionis Regio; Argyre I on the following limb is brighter. The south polar cap is bordered by a broad dusky tract, and is indented at around the CM longitude; the cap appears brighter towards the following limb. 300 mm Newtonian, 260×, integrated light (Credit: Paul Stephens)

Chapter 9

Recording Mars

9.1 Visual Observation

In order to see the finer detail on Mars, good weather conditions on both Mars and Earth is needed! From the Earth's surface we're looking through a thick layer of atmosphere – 99% of it lies in a layer just 31 km thick. The lower Mars appears in the sky, the more atmosphere its light must travel through to reach the observer's keen eye, its light diminishing and levels of turbulence generally increasing nearer the horizon.

Most problems occur in the bottom 15 km of the atmosphere. Clouds are the most obvious impediment to astronomical observation, but even a perfectly cloud-free sky can be useless for telescopic observation. Atmospheric air cells of varying sizes (2–20 cm) and density refract light slightly differently. The worst turbulence is produced when the air cells are mixing vigorously, making the light from celestial objects appear to jump around. The observer's immediate environment also plays a significant role in how good the image is. A telescope brought out into the field needs some time to cool down. Chimneys, houses and factory roofs that give off heat produce columns of warm air that mix with the cold night air to warp the image.

There's no substitute for altitude. To maximize the view of Mars on any given evening, it's best to observe when the planet is as high as possible, local conditions permitting. Mars can reach an opposition declination of 25° in Taurus, as on 8 December 2022 – a transit altitude of 64° from London and 74° from New York, but only 31° from Sydney. The lowest Martian declination at opposition in coming years occurs on 26 July 2018, when the planet will be −25° in Capricornus – a transit altitude of just 14° from London and 24° from New York, but 81° from Sydney.

Seeing varies from 0.5 arcsecond resolution on an excellent night at a world-class observatory site, to 10 arcseconds on the worst nights. On nights of poor seeing it's hardly worth observing Mars with anything but the lowest powers, since turbulence in the Earth's atmosphere will make the planet appear to shimmer, rendering any fine detail impossible to discern. For most of us, seeing rarely allows us to resolve detail finer than 1 arcsecond, regardless of the size of the telescope used, and more often than not a 150 mm telescope will show as much detail as a 300 mm telescope, which has a light gathering area four times as great. It is only on nights

P. Grego, *Mars and How to Observe It*, Astronomers' Observing Guides,
DOI 10.1007/978-1-4614-2302-7_9, © Springer Science+Business Media New York 2012

of really good seeing that the benefits of the resolving power of large telescopes can be experienced – such conditions don't occur frequently enough! Seeing scales are discussed below.

9.2 Pencil Sketches

Unlike many other branches of Solar System astronomy, the competent visual observer can still produce observations that show detail approaching the best amateur CCD images. Visual observers can make the most out of mediocre seeing conditions, teasing out finer planetary detail during those moments of improved seeing.

Observational drawings of Mars are usually made on 50 mm diameter circular blanks, but the ALPO Mars Section uses a 42 mm blank (the reasoning: Mars is 4,200 miles in diameter) but this tends to be a little too small for an entirely satisfactory representation. Before observing, it is advisable to indicate the planet's phase and the orientation of its axis on the outline blank, rather than having to guess this at the eyepiece. Tabular information given in various detailed Martian ephemerides will allow an accurate blank to be constructed; to do this, the *BAA Handbook* gives the following relevant information: P = Position angle of the north pole, measured in degrees east from the north point of disk; Q = Position angle of the point of greatest defect of illumination (the point where the phase is at its greatest width), measured east in degrees from the north point of the disk; Phase (percentage or fraction of disk illuminated); Tilt = Tilt of Mars' north pole, in degrees, towards (positive value) or away (negative value) from the Earth. This information is given in intervals of 10 days, so some interpolation of the data is usually required. Numerous astronomical computer programs will display a graphic showing an accurate latitude and longitude grid superimposed upon an image of Mars, along with the other essential information.

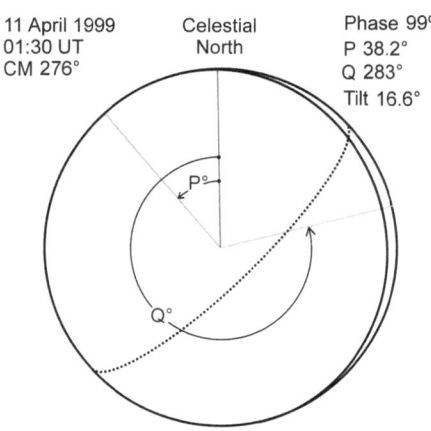
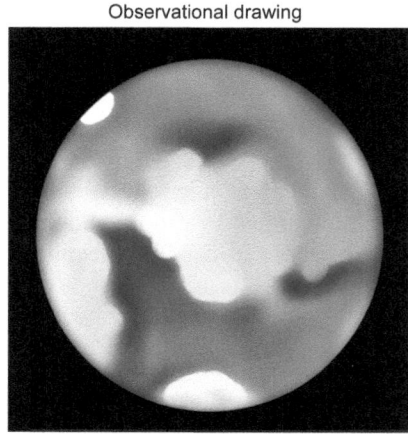

Construction of a Mars observing blank (Credit: Grego)

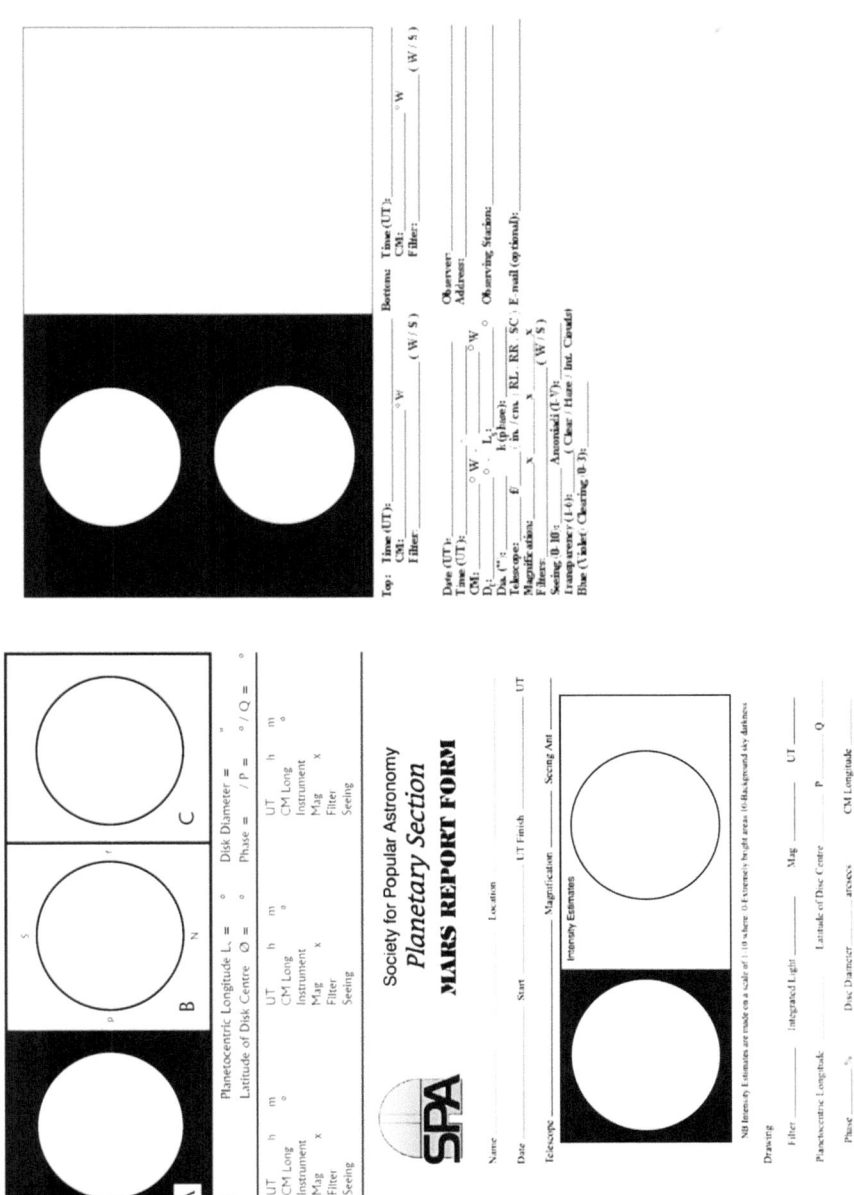

Mars observing forms produced by the BAA Mars Section, SPA Planetary Section and ALPO Mars Section, all available to download online (Credit: Grego)

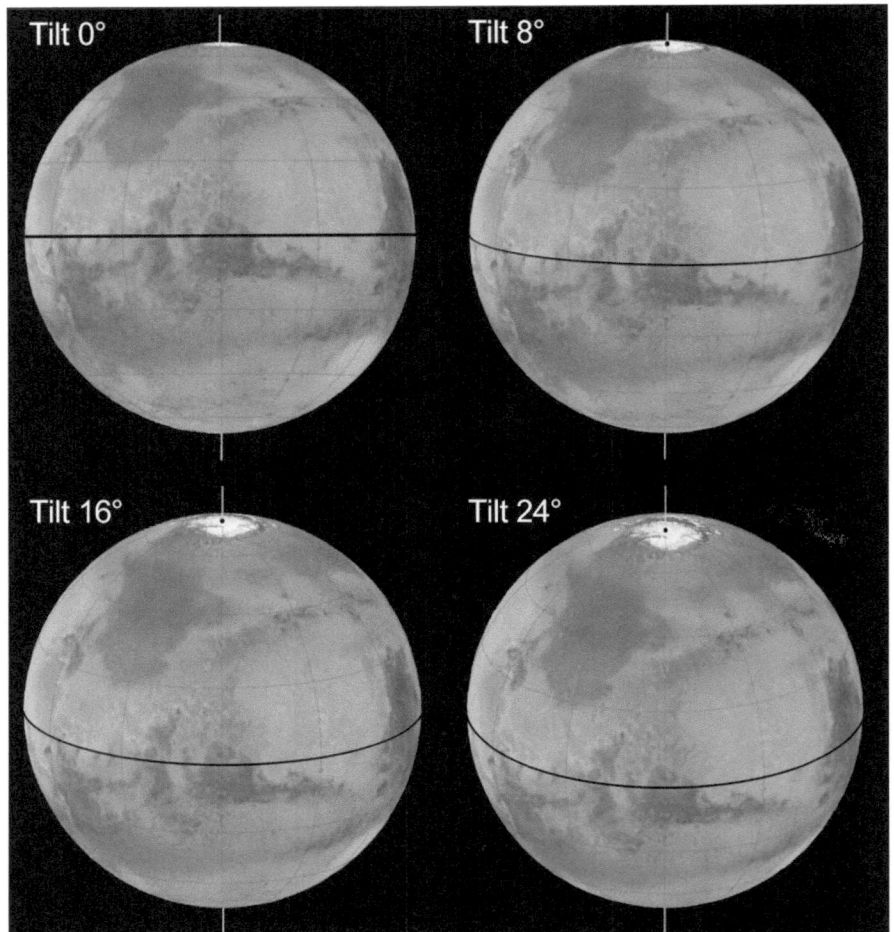

The apparent tilt of Mars, favouring the planet's northern hemisphere (Credit: Grego)

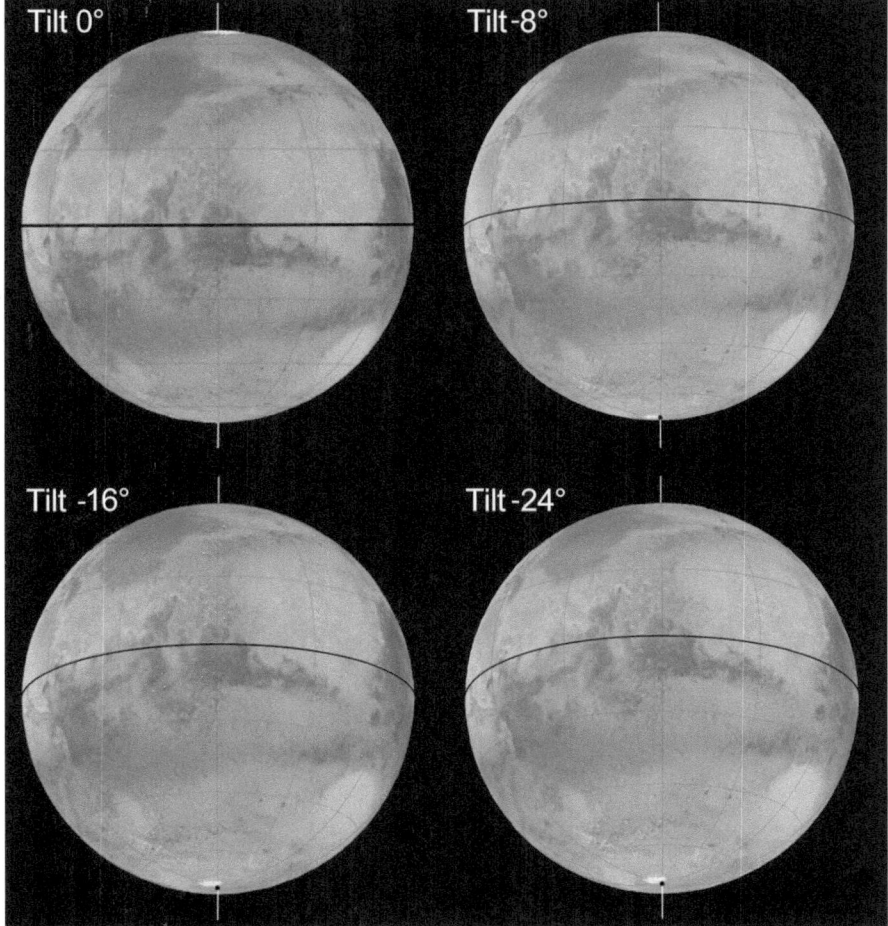

Tilt 0° Tilt -8°

Tilt -16° Tilt -24°

The apparent tilt of Mars, favouring the planet's southern hemisphere (Credit: Grego)

A prior knowledge of the planet's central meridian during the observing session is also a considerable advantage. Astronomical ephemerides, including the *BAA Handbook*, give data on the central meridian's position for 00 h UT each day, which can then be added to or subtracted from according to the time of the observation (see below, central meridian transit timings). Many observers avoid the mathematics by consulting a suitable computer program to display a graphic representation of the hemisphere presented during the observing session. However, it is important to bear in mind that the shapes and intensities of the features shown on the computer monitor are likely to differ in many ways from the actual view through the eyepiece, so it must be considered as a guide only.

9.3 Making the Sketch

Visual observers should attempt to depict Mars as accurately and as unambiguously as possible. The drawings need not be great masterpieces of art, and no observer's work need be considered of more value than another's because it simply looks nicer. The aim is to attend to the fine detail visible through the eyepiece and record it to the best of your abilities. The finished results will be the products of your efforts and a permanent record of your planetary forays. Don't throw them away – keep all your observations in a folder and you may be pleasantly surprised at how your skills of planetary depiction improve over time.

Lack of preparation can often lead to disappointing results. It's not uncommon to see the would-be planetary observer fumbling around at the eyepiece – pencil in one hand, sketchpad in the other, flashlight in mouth – attempting to depict the outline of a planet by freehand or by drawing around the edge of an eyepiece barrel.

Use a standard BAA, SPA or ALPO Mars observing form or prepare for yourself a phase-corrected outline blank; if intensity estimates are to be made in addition to a pencil sketch, two equal blanks should be drawn side by side on the same piece of paper; this will minimize having to write down identical details on two sheets, and it will avoid confusion if the sketch and intensity estimate become separated.

When printing off report forms I use a monochrome laser printer and 100 gsm white paper, because inkjet prints have an annoying habit of smudging once they become damp.

A set of soft-leaded pencils, from HB to 5B is recommended. It is reasonable to set yourself about half an hour per observational drawing. Patience is essential – even if the clouds are threatening to obscure the planet from view, or if your fingers are feeling numb with cold – because a rushed sketch is bound to be less accurate.

Don't expect to see much in the way of detail on Mars if you hastily set up your telescope and begin observing straight away. If you're taking your telescope from a warm indoor environment to a colder outdoors, allow your telescope a good while to acclimate and cool down. This is especially important with sealed-tube instruments like SCTs and MCTs, otherwise you'll get a poor image caused by turbulent air currents, and the observation will hardly be worth making.

An obvious starting point for an observational drawing is the planet's bright polar cap. Next, the outlines of the more prominent darker features can be lightly sketched in, along with any particularly well-defined bright features; the use of a soft pencil gives you the chance to erase anything if the need arises. It is best not to depict the dark sky around a planet. Fine detail and more distinct tonal shading should be left until last, once the outlines of the main features have been decided upon. You're unlikely to see any really dark features, but any unusually dark areas should be drawn by applying minimal pressure on the paper using layers of soft pencil rather than a single layer made with heavy pencil pressure. Soft graphite smudges really well, and good use of smudging along the terminator or along the edges of cloud features can produce smooth blends.

No drawing of Mars, however detailed, is likely to convey a visual impression fully, so many observers make notes to accompany their sketches, say to mention any obscure or uncertain features that may have been observed but not adequately depicted, in addition, of course, to the vital observing data which ought to accompany every observational drawing.

9.4 Cybersketching

The use of graphite pencil means that the Red Planet has traditionally been depicted in greyscale; it is, after all, quite difficult to make colored pencil drawings (or renditions in pastels or paint) while seated in the dark at the telescope eyepiece. Thankfully, new technology has come to the observer's aid – cybersketching with a portable computer allows colors to be added to an observational sketch almost as easily as it is to make a greyscale drawing.

In this context, portable computer means any standalone touch-screen device small enough to be used comfortably at the telescope eyepiece; this definition covers PDAs (personal digital assistants), tablet computers (operating on a variety of systems, from Windows to Android), iPhone and iPad. These can all be operated by a stylus of one sort or another, which takes the place of a pencil. It must be pointed out, however, that styli for the finger-touch screens of the iPhone and iPad require a large tip with a broad area of contact, so fine control of a drawing is somewhat more problematic than the sleeker styli used by PDAs.

Portable computers and drawing programs have a variety of advantages over traditional pencil sketches. With their adjustable backlit screens, there's no need for an additional torch to illuminate the drawing; there's no more damp paper and unwanted smudges either. An impressive range of drawing tools can be found within such programs, such as the ability to draw accurate phase-corrected blanks and a variety of media effects, nib sizes, smear, blur, blend and color adjustment, to name but a few, makes producing observational drawings an enjoyable process.

A number of excellent drawing programs are available for portable computers. An old but trusty drawing program that I often use on my equally old but trusty PDA (SPV M2000, Windows Mobile 2003 SE OS) is Mobile Atelier; this saves images in 8-bit BMP format at a resolution of 240 × 320 pixels. This degree of resolution is okay for general astronomical cybersketching at the eyepiece, including disk drawings of Mars. More detailed drawings are capable of being produced in other programs, one of my favorites being the very versatile Pocket Artist 3.

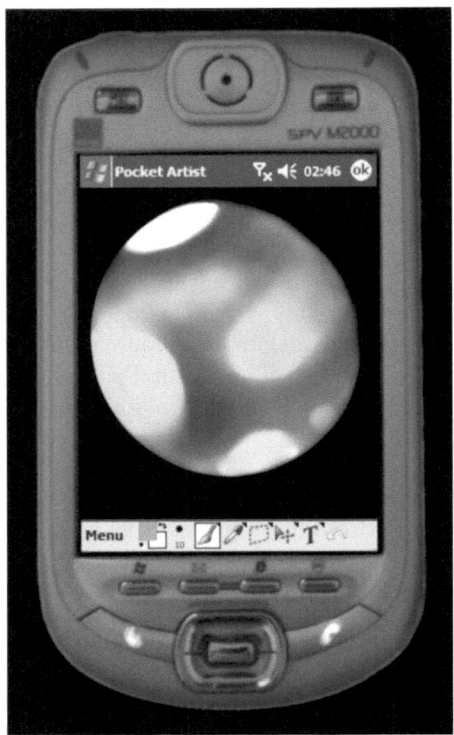

The author's venerable SPV M2000 PDA, showing a typical observational cybersketch of Mars made at the eyepiece (Credit: Grego)

9.5 Line Drawings and Intensity Estimates

Outline drawings may be an alternative or adjunct to making shaded pencil drawings at the eyepiece. Annotated line drawings can be every bit as informative as a shaded pencil sketch, but the technique should not be considered a quick and easy alternative to tonal drawing, since they ought to be drawn just as carefully and with the same amount of attention to detail. Intensity estimates are a valuable addition to line drawings. This requires the observer to estimate of the brightness of each distinct area depicted on the drawing, using a scale of 0–10, with 0 representing the brilliant white polar cap, 10 the black night sky. An intensity estimate of 2 might be allocated to Mars' bright desert regions, and particularly dusky features may appear as dark as 7.

Mars' surface features may be subject to temporary seasonal changes, or changes taking place over a number of apparitions, so an indication of the apparent intensities of features is a useful way of monitoring these changes. Seasoned users of intensity estimates usually become so adept at the technique that fractions of points are used. Intensity estimates ought to be performed when observing in integrated light alone, without a filter.

The eye is capable of differentiating between hundreds of shades of grey, so a skilled observer can easily subdivide the basic scale yet further. Unlike, say variable star estimations, these tend to be qualitative visual estimates, rather than quantitative ones. Additionally, optical illusions are capable of playing tricks with the observer – two areas of true identical tone may appear to have widely different tones when set next to areas of different brightness and contrast.

9.6 Copying up Your Observations

Few people have the ability to produce completely error-free observational drawings at the eyepiece. Therefore it is best to prepare a neat copy of your observational drawing as soon after the observing session as practically possible, while the scene remains vivid in your mind, since accurate recall tends to fade pretty quickly. A fresh drawing prepared indoors will be far more accurate than the original telescopic sketch that it is based upon, as the observer will be able to recall little things about the original sketch that perhaps weren't quite right and needed to be rectified on the neat drawing. Unlike capturing an image on CCD, many of the details of the observation are contained inside the observer's head and are accessible only using the mind's eye.

It is perfectly possible to translate the most rough and ready drawing made at the eyepiece into a far more accurate and pleasing depiction of what was observed. Some of my own at-the-eyepiece observational drawings (a few, I'm embarrassed to admit, were made freehand and in ballpoint pen on lined paper owing to a lack of resources while in the field) are pretty worthless in themselves, inaccurate and quite horrible to look at. However, their real value is a transient thing, accessible only to the observer a few hours after the observing session and only capable of being revealed in a fresh, neat observational drawing. The neat copy of your original telescopic sketch can be used as the template for further drawings, or used to electronically scan or photocopy.

Copied drawings may be prepared in a variety of media. Superb results can be achieved using India ink washes, and paintings in gouache or acrylics are excellent for reproducing observation on a larger scale for exhibition purposes. Both these techniques require proficiency in brushwork, and although a description of the methods involved is beyond this book, practice, experimentation and perseverance will pay big dividends.

When making physical drawings, soft pencil on smooth white paper is by far the quickest and least fussy medium. Once completed, pencil drawings need to be sprayed with a fixative so that they don't smudge if they are inadvertently rubbed. Regular copy shop photocopies of tonal pencil drawings are not quite good enough to submit to astronomical society observing sections or for publication in magazines, as the full range of tones in the drawing will not be captured, and it may appear somewhat dark and grainy. These days, however, most planetary observing sections are happy to accept a high quality laser print or a digitally scanned drawing. Some commercial magazines may insist on having the original artwork to work from, or at the very least for high quality, high resolution scans to be submitted on floppy disk or by email.

Tempting though it may be to discard old drawings, they represent a permanent record of your observations and hard work at the eyepiece. At the very least, comparing old observations with more recent ones will demonstrate how much your skills of observation and recording have improved. Original drawings can be used as the basis for subsequent copies for the observing sections of any astronomical societies to which you belong, or for publication. For these reasons, do hang on to your original drawings for future reference. Devote a folder or a ring binder to your original drawings of Mars, placing each in a clear polythene pocket, and store it in a clean, dry environment.

9.7 Written Notes

All observational drawings should be accompanied by notes, made on the same sheet of paper, stating the name of the planet (to avoid any possible confusion with Venus or Mercury – and this can happen), along with the date, observation start and finish times (in Universal Time), the instrument and magnification employed, integrated light or filters used, and the seeing conditions. Short written notes, made at the telescope eyepiece, may also point out any unusual or interesting features that have been observed or suspected, but which may not necessarily be obvious or able to be depicted on your drawing.

Date and time: Amateur astronomers around the world use UT (Universal Time), which is the same as GMT (Greenwich Mean Time). Observers need be aware of the time difference introduced by the world time zone in which they reside and any local daylight savings adjustments to the time, and convert this to UT accordingly – the date should be adjusted too. Times are usually given in terms of a 24-h clock – for example, 3.25 pm UT can be written as either 15:25, 1525 or 15 h 25 m UT.

Seeing: To estimate the quality of astronomical seeing, astronomers refer to one of two scales of seeing. In the UK, many observers use the Antoniadi Scale, devised specifically for lunar and planetary observers:

AI – Perfect seeing, without a quiver. Maximum magnification can be used if desired.

AII – Good seeing. Slight undulations, with moments of calm lasting several seconds.

AIII – Moderate seeing, with large atmospheric tremors.

AIV – Poor seeing, with constant troublesome undulations.

AV – Very bad seeing. Image extremely unstable, hardly worth attempting to observe, since even the planet's phase may not be able to be discerned.

In the United States, seeing is often measured from 1 to 10 on the Pickering Scale. The scale was devised according to the appearance of a highly magnified star and its surrounding Airy pattern through a small refractor. The Airy pattern, an artefact introduced by optics, will distort according to the degree of atmospheric turbulence along its light path. Under perfect seeing conditions, stars look like a tiny bright point surrounded by a complete set of perfect rings in constant view. Of course, most planetary observers don't check the Airy pattern of stars each time they estimate the quality of seeing during an observing session – an estimate is made, based on the steadiness of a bright stellar image.

P1 – Terrible seeing. Star image is usually about twice the diameter of the third diffraction ring (if the ring could be seen).

P2 – Extremely poor seeing. Image occasionally twice the diameter of the third ring.

P3 – Very poor seeing. Image about the same diameter as the third ring and brighter at the center.

P4 – Poor seeing. The central disk often visible; arcs of diffraction rings sometimes seen.

P5 – Moderate seeing. Disk always visible; arcs frequently seen.

P6 – Moderate to good seeing. Disk always visible; short arcs constantly seen.

P7 – Good seeing. Disk sometimes sharply defined; rings seen as long arcs or complete circles.

P8 – Very good seeing. Disk always sharply defined; rings as long arcs or complete but in motion.

P9 – Excellent seeing. Inner ring stationary. Outer rings momentarily stationary.

P10 – Perfect seeing. Complete diffraction pattern is stationary.

Considerable confusion can be caused if a simple figure is used to estimate seeing, without indicating whether it's made on the Antoniadi or Pickering scale. So, in addition to designating the seeing with a letter and a figure (AI–V or P1–10), a brief written description of seeing, such as 'AII – Good with occasional moments of excellent seeing' can be made.

Conditions: An indication of the prevailing weather conditions, such as the amount of cloud cover, the degree and direction of wind and the temperature.

Transparency: The quality of atmospheric clarity, known as transparency, varies with the amount of smoke and dust particles in the atmosphere, along with cloud and haze. Industrial and domestic pollution causes transparency to be worse in and around cities. A transparency scale of 1–6 is often used, according to the magnitude of the faintest star detectable with the unaided eye.

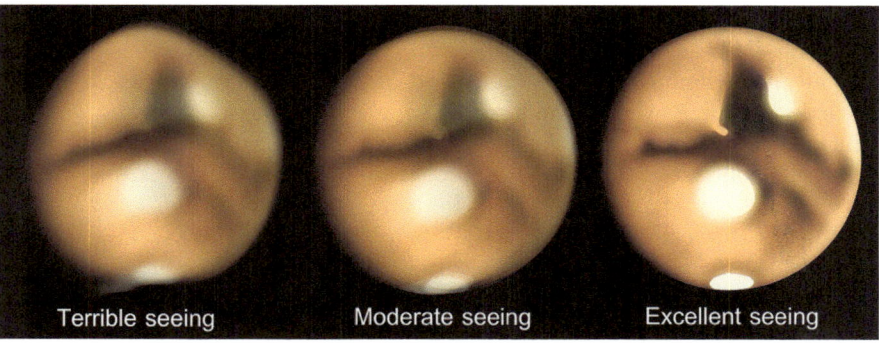

Seeing affects the quality of the image in the eyepiece. Shown here are simulations of terrible seeing (not worth making a sketch), moderate seeing and excellent seeing (Credit: Grego)

9.8 Mars Data

Much of the necessary information about Mars is published in annual astronomical ephemerides such as the *BAA Handbook* and the *Multiyear Interactive Computer Almanac* (MICA, published by the US Naval Observatory). Many computer programs and online resources also deliver this important information. Typical data for Mars given in astronomical ephemerides includes:

Date of opposition and conjunction with the Sun.

Date (given in equally-spaced increments, e.g., 5 day intervals).

Planet's celestial co-ordinates in RA (Right Ascension) and Dec (Declination).

Magnitude (to nearest 1/10 of magnitude).

Diameter (in arcseconds).

Phase (given as a percentage or as a decimal number, e.g., 97% or 0.97).

Elongation (in degrees, E expressed as a positive and W as a negative value).

Central Meridian (CM). The Martian longitude, given at daily intervals for 00 h UT

Distance from the Earth in Astronomical Units (AU).

9.9 Filter Work

Integrated light observations reveal a great deal of Mars' surface features and its atmospheric phenomena, but the use of filters enables a much more intensive scrutiny of the red planet's markings and atmosphere. Importantly, filter work requires the image produced in a given light to be reasonably bright – it's pointless to attempt to observe using a filter which gives a very dim image.

The most commonly used filter types are red (Wratten 25), orange (Wratten 21) and yellow (Wratten 15). Red, orange and yellow filters make the dark surface features easier to see and will improve the definition of less distinct markings in brighter desert regions. Yellow colored dust storms which blow up on Mars from time to time appear brighter through these filters, especially in red light. Yellow (Wratten 21) and green (Wratten 58) filters improve the visibility of low altitude orographic clouds.

Atmospheric limb brightening, haze and high altitude orographic clouds are revealed using blue (Wratten 44A and 80A) and blue-violet (Wratten 47) filters, while the relative brilliance of the polar regions is maintained but at the same time muting the contrast of the dusky surface features. A phenomenon known as the 'blue clearing' occasionally takes place in the Martian atmosphere, when, using a blue filter (observers use the Wratten 47 as a standard for this) surface features become easier to see, perhaps over a period of several days; its precise cause is poorly understood. At least a 200 mm telescope is required to effectively use the Wratten 47 filter, because of its low light transmission.

Filters suitable for visual observation: *Red* (W25), *Yellow* (W15), *blue* (W44A) and *green* (W58) (Credit: Grego)

To assess the state of the Martian atmosphere and the intensity of the blue clearing, the following scale is used:

0. No dark surface features are discernable
1. Some dark surface features are vaguely visible
2. Dark surface features are easy to see
3. Surface features are almost as clearly defined as in integrated light (very rare)

9.10 Central Meridian Transit Timings

The imaginary line running down the center of a planet's disk, joining both poles, is known as the central meridian. Timings of features crossing this line have in the past been a valuable means of mapping planets, particularly Mars, Jupiter and Saturn. Making accurate timings of the passage of Martian features across the planet's central meridian is hampered by the fact that Mars rotates much more slowly than Jupiter and Saturn and more often than not it displays a phase, making it difficult to judge the position of the central meridian.

When either of Mars' poles is tilted strongly towards the Earth, features appear to make a markedly curved path across the disk as the planet rotates. Prior to opposition, when the planet's following limb is fully illuminated, a prominent feature will appear near the following limb and proceed to transit the planet's central meridian in around five and a half hours; this feature will disappear into the planet's evening terminator at the preceding limb in less time than it took to reach the meridian – the exact time period depends on the planet's phase. At around opposition, the planet is fully illuminated and both preceding and following limbs are clearly defined. After opposition, the following limb is encroached upon by the morning terminator, and features will reach the central meridian from the terminator in a far shorter period than they take to move from the central meridian to disappear near the fully illuminated preceding limb.

Although Mars rotates on its axis more than twice as slowly as Jupiter, it is possible to estimate a Martian feature's central meridian passage within a few minutes of accuracy. Jupiter has the advantage of showing less of a phase and being almost squarely-presented to the Earth; since its belts and zones appear as more or less straight lines, features simply appear to move from one limb to the other along a straight line; the position of Jupiter's central meridian is therefore rather easy to picture in the mind's eye. Although one point along Mars' central meridian can usually be judged from the position of whichever bright polar cap is presented, it is always helpful to consult a detailed planetary ephemeris or a computer program to calculate the precise orientation of the planet's axis and the angle of its central meridian.

While central meridian transit timings of regular Martian features may not be of immense scientific value, they are enjoyable to make, and will enable the observer to draw up an independent map of the planet's features based on calculations of the planet's longitude given in astronomical ephemerides and computer programs. Transit timings can be of scientific value in pinpointing the location of any unusual Martian features, such as bright orographic clouds or localized dust storms, or any unusual transient phenomena. Importantly, the observer ought never assume that his or her observations are pointless because someone else with a bigger telescope and better equipment is doing the observing at the same moment. Many major planetary discoveries have been made by keen amateurs prepared to observe and make records of Solar System objects during periods when most other observers are likely to be neglecting that particular object near the beginning or end of its apparition.

9.11 Change of Martian Longitude in Intervals of Mean Time

h	°	h	°
1	14.6	6	87.7
2	29.2	7	102.3
3	43.9	8	117.0
4	58.5	9	131.6
5	73.1	10	146.2

m	°	m	°	m	°
10	2.4	1	0.2	6	1.5
20	4.9	2	0.5	7	1.7
30	7.3	3	0.7	8	1.9
40	9.7	4	1.0	9	2.2
50	12.2	5	1.2	10	2.4

Example

On 28 August 2003, the northern tip of Syrtis Major was noted cross the central meridian of Mars at 22:18 UT. The ephemeris gives a figure of longitude 320.5° for Mars' central meridian at 00 h on 29 August. The difference between 22:18 and 00 h is 1 h 42 m. Converting this to longitude: 1 h = 14.6° + 40 m = 9.7° + 2 m = 0.5° − Total = 24.8°. Deducting 24.8° from 320.5° gives a longitude of 295.7° for the northern tip of Syrtis Major, as observed on that evening.

9.11.1 Imaging Mars

Traditional photography has never been able to show as much detail on Mars as a drawing made at the telescope eyepiece under identical conditions. Until CCD imaging became a powerful tool in amateur astronomers' hands in the last decade of the twentieth century, drawing at the telescope eyepiece was the only means of recording fine detail and subtle shadings on Mars.

Few professional observatories ever turn their big telescopes towards Mars – other than perhaps to impress visitors and students or to test out new equipment. Of course, many amateur astronomers never consider undertaking systematic studies, but instead choose occasionally to view Mars for the sheer challenge and the visual pleasure that they provide.

Imaging and observational drawing of Mars are practices which have been entirely undertaken by amateur astronomers for many decades. Happily, Mars offers plenty of interest to occupy the observer, and indeed, the dedicated amateur stands a good chance of discovering transient Martian atmospheric phenomena.

9.12 Conventional Photography

A beautiful bright, pin-sharp image of Mars can often be viewed using a small telescope, so it may seem reasonable to suppose that the camera is capable of capturing the same image with little difficulty. Mars is often bright enough to register on just about any conventional film camera pointed through the eyepiece, but successful planetary photography using an ordinary film camera – be it a simple compact camera or a 35 mm SLR (single lens reflex) – can be a rather difficult and involved process. Digital photography has now rendered film photography obsolete, but there still remains a die-hard core of imagers who still use the old technology to capture celestial objects.

An equatorially driven telescope and a firmly fixed camera, keeping the planet steadily centered in the field of view, are essential to capture Mars on conventional film. While visual observers can contend with the movement of a planet as it drifts through the field of view of an undriven telescope, the slightest movement will produce motion blurring in a photographic image; the longer the exposure, the greater the degree of blurring.

9.13 Afocal Photography

Afocal photography is the process of photographing an object through the telescope eyepiece with a non-SLR compact film or digital camera. Basic compact film and digital cameras have fixed lenses that are usually preset to focus objects ranging from a few meters away to infinity; with their preset exposures for standard photography, they are unable to be fine-tuned for astronomical purposes. They are however suitable for imaging a bright planet like Mars.

Non-SLR film cameras and some compact digicams have a viewfinder that is slightly to one side of the photographic optical axis – fine for composing quick and easy images of everyday scenes, but completely unusable for afocal photography, as it will not show the image of Mars being projected into the camera. Given that the view through the camera will not be visible during afocal photography with regular film cameras, Mars must be lined up in the crosshairs of an accurately aligned finderscope and photographed without the imager being able to see the picture's composition.

Afocal photography through most compact digital cameras is somewhat easier because the image being projected onto the camera's CCD chip appears on an electronic display – albeit a somewhat coarse and small LCD screen – at the rear of the camera. Digital cameras are designed for everyday use, so their automatic settings may pose considerable problems when attempting to photograph Mars, so experimentation with the camera's various settings is necessary to produce the best results.

First, Mars is focused by eye through the telescope eyepiece, and the camera is then positioned close to the eyepiece and held there firmly. If there is a focus adjustment on the camera, it's best set to infinity. Some basic cameras are meant to be used without any accessories, so some kind of makeshift camera adapter will need to be constructed to mate it to the telescope – many cameras are lightweight enough to be temporarily fixed to a telescope with a little blu-tack and electrical tape. If your camera has a standard tripod bush at the bottom of the body, it will be possible to mount the camera more securely or attach it to a commercially available telescope camera mount.

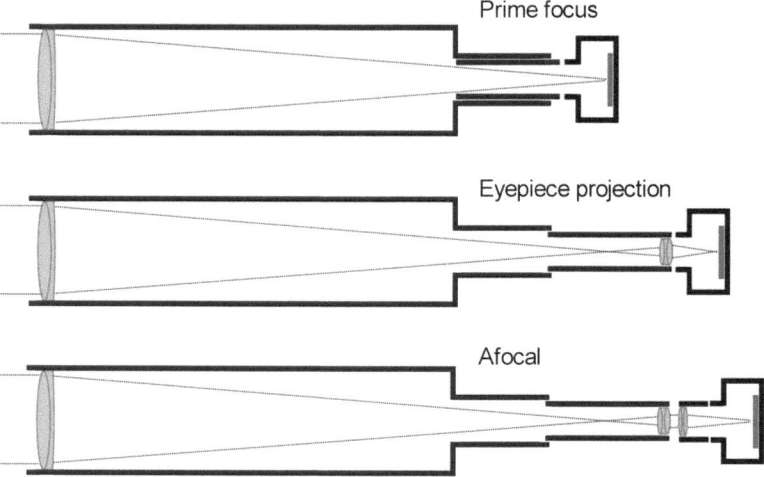

Three imaging techniques – prime focus, eyepiece projection and afocal imaging (Credit: Grego)

9.14 SLR Film Photography

SLR cameras direct the light from the object being photographed through the main camera lens to the eye via a mirror, prism and eyepiece. The area framed in the viewfinder is exactly the area that will be included in the final image. When the shutter button is pressed, the mirror instantly flips out of the way of the light path, allowing it to project directly onto the film.

9.15 Film Types

A film's ISO rating indicates its 'speed' – the higher the ISO, the faster the film and less exposure time is required. 200 ISO is a medium speed film, and inexpensive generic 200 ISO color print film is fine for beginning planetary photographers to experiment with. Quality tends to vary from brand to brand, and even varies among different batches of the same 'budget' brand. Slower films have finer grains allowing more detail to be captured, while grain size increases with a film's ISO rating. This may not be evident in comparing regular sized photographic prints, but enlargements will clearly show the difference. Photographs taken with slower film can withstand much more enlargement than those taken with faster film, but slow films also have the drawback of requiring more time to expose. High magnification shots of the planets made on regular film require an accurately driven equatorial drive.

Some of the most stunning images of Mars are wide angle twilight views taken with SLR cameras piggybacked on a telescope or even mounted on an undriven tripod. The composition of such images is very much a matter of taste, but there are a few basic rules to taking a wide-angle image of sufficient interest and spectacle to merit its inclusion on the Astronomy Picture of the Day website (URL in appendix). First, include an interesting foreground – say, a historical site, an area of natural beauty or a panoramic cityscape. Reflections of bright celestial objects like Mars from bodies of water set off a picture nicely. Events such as close approaches of Mars to the Moon, other planets and deep sky objects make an image really memorable, so keep a look out for suitable photographic opportunities by scanning the sky diary pages of your favourite astronomy magazine or by running your PC astronomy program.

9.16 Prime Focus Photography

When a camera body (a camera minus its lens) is attached to a telescope (minus its eyepiece), the light falling onto the CCD chip or film is at the main focal point of the telescope's objective lens (or mirror). Conventional prime focus photography, even when made with a long focal length telescope, delivers a small planetary image (tiny on 35 mm film). A Barlow lens (standard ones come in x2 and x3 varieties) will effectively increase the focal length of a telescope, producing an enlarged image. Focusing a planet at prime focus is done by looking through the camera's viewfinder, using the telescope's own focuser. The process is usually quite forgiving, and by no means as exacting as focusing an image using eyepiece projection (see below).

Prime focus planetary imaging is extensively used with webcams and astronomical CCD cameras. CCD chips are tiny in comparison to 35 mm film, so the effective magnification they deliver is far larger (see below for CCD photography). Precise focusing is critical, and best achieved using an electronic focuser rather than manual knob-twiddling.

9.17 Eyepiece Projection

High magnification planetary photographs can be achieved by inserting an eyepiece into the telescope and then projecting the image into the camera, minus its lens. Adapters are widely available that fit into standard sized eyepiece holders (1.25-in. and 2-in. diameters) and the bodies of various makes of SLR. Orthoscopic and Plössl eyepieces deliver crisp images with flat fields of view. Eyepiece projection will deliver a far higher magnification image than prime focus photography, the degree of magnification depending on the focal length of the telescope and the eyepiece, and the distance of the eyepiece from the CCD or film plane. Shorter focal length eyepieces will deliver higher magnifications, and increasing the distance of the eyepiece from the CCD or film plane will also increase magnification. Focusing is achieved by looking directly at the magnified planetary image through the camera viewfinder and adjusting the telescope's focusing knob until the planet appears sharply focused.

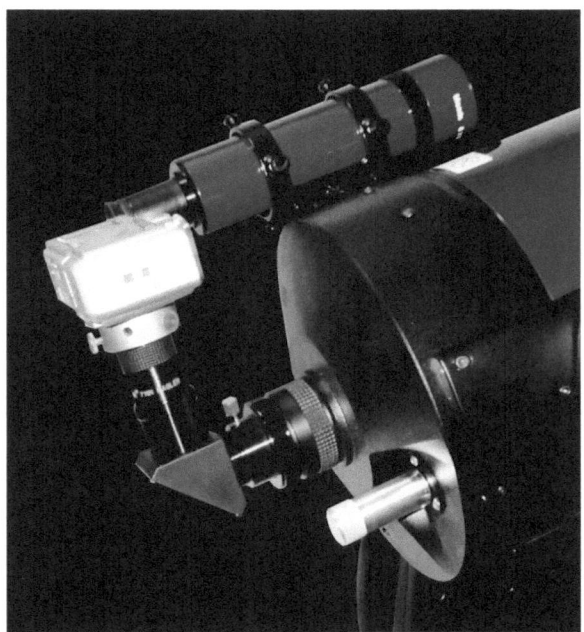

A digital camera set up for afocal imaging with a 200 mm SCT (Credit: Grego)

9.18   Digital Imaging

A CCD (charge coupled device) is a small flat chip – about the diameter of a match head in most commercial digicams – made up of an array of tiny light sensitive elements called pixels. Light hitting each pixel is converted to an electrical signal, and the intensity of this signal directly corresponds to the brightness of the light striking it. This information can be stored digitally in the camera's own memory or transferred to a PC, where it can be processed into an image. CCDs in the lowest-end 0.3 megapixel digital cameras have a lowly array of 640×480 pixels, while a 10 megapixel camera will boast a $3,872 \times 2,592$ pixel chip.

Webcams and dedicated astronomical CCD cameras enable amateur astronomers with quite modest equipment the opportunity to obtain very satisfying images of Mars. Digital images are infinitely easier to enhance and manipulate than a conventional photograph in a photo lab darkroom. Although just about anyone can take an acceptable planetary snapshot by pointing a CCD camera through a small telescope, it requires considerable skill and expertise – both in the field, and later at the computer – to produce high resolution images that show surface detail.

9.19   Camcorders and Vidcams

Camcorders have fixed lenses, and footage must be obtained afocally through the telescope eyepiece. The same problems that affect afocal imaging using conventional film and digital cameras apply to camcorders. Camcorders tend to be heavier than digital cameras, and it is essential that the camcorder is coupled to the telescope eyepiece as sturdily as possible. Some of the same equipment designed to hold a digital camera in place when taking afocal planetary images can be used to secure a lightweight camcorder to a telescope.

Digital camcorders are the lightest and most versatile camcorders, and their images can be easily transferred to a computer for digital editing using the same techniques as images obtained with a webcam (see below). Once downloaded onto a computer, individual frames from digital video footage can be sampled individually (at low resolution), stacked using special software to produce detailed, high resolution images, or assembled into clips that can be transferred to a CD-ROM, DVD or videotape. The process can be time consuming – the time spent running through the video footage and processing the images may amount to far longer than the time that was actually spent taking the footage. Digital video editing also consumes a great deal of a computer's resources, in both terms of memory and storage space – the faster a computer's CPUs and graphics card, the better. At least 5 gigabytes need to be available on your computer's hard drive for the most basic editing of video clips.

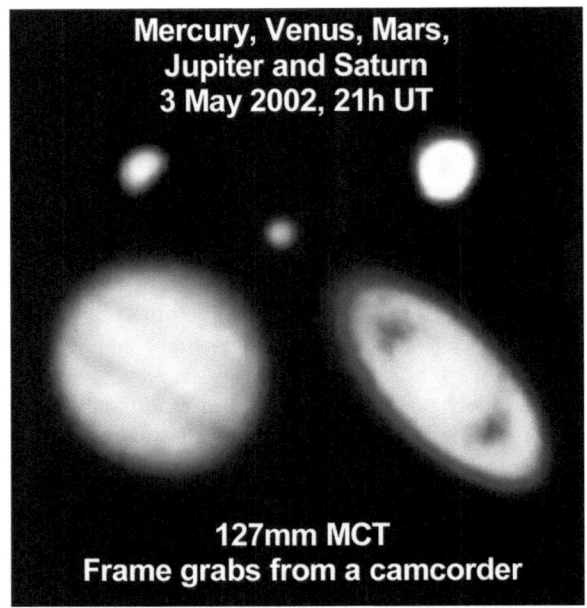

Using a 127 mm MCT and a camcorder, the author observed and captured all five classical planets within just a few minutes of each other on the memorable evening of 3 May 2002 — Mercury (upper left), Venus (upper right), Mars (center), Jupiter and Saturn. The images aren't particularly good, being simple digital frame grabs (unstacked), but all are shown to scale (Credit: Grego)

9.20 Webcams

Although they are usually designed for use in the home to enable communication between individuals over the Internet, webcams can be used to capture high resolution images of the planets. Lightweight and versatile, webcams are just a fraction of the cost of dedicated astronomical CCD cameras. Just about any commercial webcam hooked up to a computer and a telescope can be used to image Mercury and Venus, although the quality of the images depends on a number of skills and techniques that may only be improved through patience, practice and perseverance.

While webcams may not have as sensitive a CCD as more expensive astronomical CCD cameras, their ability to record video clips made up of hundreds, or thousands of individual images gives them a distinct advantage over the single-shot astronomical CCD. By taking a video sequence made up of dozens, hundreds or even thousands of individual frames, the effects of poor seeing can be overcome by processing only the clearest images in the video clip. These images – selected either manually or automatically – can then be combined using stacking software to produce a highly detailed image. This may show as much detail as that seen visually through the eyepiece using the same instrument.

Webcams are usually used at the telescope's prime focus (minus the webcam's original lens) to image Mercury and Venus. A number of webcams, such as the popular Philips ToUcam Pro, have easily removable lenses; commercially available telescope adapters can be screwed in their place, permitting easy attachment to a telescope. Some webcams however require disassembly to remove the lens, and the adapter needs to be home made.

CCD chips are sensitive to infrared light, and the original lens assembly usually contains an infrared blocking filter – without the filter, a really clean focus is not possible, since infrared is focused differently to visible light. IR blocking filters are however available to fit into the telescope adapter, allowing only visible light wavelengths to pass through to a sharp focus.

Like the CCD chips in most other digital devices, webcam CCDs are very small. Used at the telescope's prime focus, the magnification produced on a webcam is usually augmented with a Barlow lens. Because of the high magnifications produced by a webcam at prime focus on an average amateur telescope, a well polar-aligned driven equatorial telescope with electric slow motion controls is essential in order to capture a relatively static video clip lasting 10 or 20 s. If the image drifts too much during the clip, the software used to process the video clip may not be able to produce a good alignment.

Focusing a webcam accurately can prove time consuming, but to achieve a rough focus, it is best to set up during the daytime and focus on a distant terrestrial object using the telescope and webcam, viewing the laptop monitor and adjusting the focus manually. Once the terrestrial object has been focused, lock the focus or mark the focusing barrel with a CD marker pen.

During the imaging session, the planet is centered in the field of view using the telescope's finderscope, and if it is accurately aligned the planet will appear on the computer screen, probably in need of further focusing. When the telescope's focus is adjusted manually, care must be taken not to nudge the instrument too hard, as the planet under scrutiny may disappear altogether out of the small field of view. Patient trial and error will eventually produce a reasonably sharp focus – once achieved, lock the focuser and mark the focusing tube's position so that an approximately sharp focus can be found quickly during subsequent imaging sessions.

Achieving a good focus makes the difference between a good planetary image and a great one – a fraction of a millimeter can make the difference between a good focus and a tack-sharp one. Manual focusing in the manner outlined above is exceedingly time consuming, and a perfect focus is more likely to be found by chance than trial and error. Electric focusers enable the focus to be adjusted remotely from the telescope, and deserve to be considered an essential accessory to the planetary imager. Electric focusers save much time and make a great difference to your enjoyment of imaging; importantly, they offer infinitely more control over fine focusing. A webcam attached to a computer through a high speed USB port will deliver a rapid refresh rate of the image, enabling fine focusing in real time.

Video sequences of the planets can be captured using the software supplied with the webcam. It is necessary to override most of the software's automatic controls – contrast, gain and exposure controls require adjusting to deliver an acceptable image. Many imagers prefer to use black and white recording mode, which cuts down on signal noise, takes up less hard drive space and eliminates any false color that may be produced electronically or optically.

The webcam's greatest strength is the sheer number of images that it provides in a single video capture. Single-shot dedicated astronomical CCD cameras costing ten times as much as a webcam are capable of taking just one image at a time;

while this image may have far less signal noise and a higher number of pixels than one taken with a webcam, in mediocre seeing conditions the chances that the image was taken at the precise moment of very good seeing are small. Webcams can be used even in poor seeing conditions, as a number of clearly resolved frames will be available to use in an extended video sequence. Video sequences are usually captured as AVI (Audio Video Interleave) files.

Astronomical image editing software is used to analyze the video sequence, and there are a number of very good freeware imaging programs available. Some programs are able to work directly from the AVI, and much of the process can be set up to be automatic – the software itself selects which frames are the sharpest, and these are then automatically aligned, stacked and sharpened to produce the final image. If more control is required, it is possible to individually select which images out of the sequence ought to be used – since this may require up to a 1,000 images to be visually examined, one after another, this can be a laborious process, but it often produces sharper images than those derived automatically.

Images can be further processed in image manipulation software to remove unwanted artefacts, to sharpen the image, enhance its tonal range and contrast and to bring out detail. Unsharp masking is one of the most widely used tools in astronomical imaging – almost magically, a blurred image can be brought into a sharper focus. Too much image processing and unsharp masking may produce spurious artefacts in the image's texture, shadowing, ghosting and chiaroscuro effects and a progressive loss of tonal detail. Each individual imager tends to develop their own particular methods of enhancing their raw planetary images. Incredible though it may seem, it is quite possible for the top few dozen of the world's most experienced amateur planetary imagers to tell each others' work apart because of the subtle differences displayed in the final image that have been introduced by different combinations of processing techniques.

9.21 Notes on Digital Imaging by Mike Brown

With the advent of cooled digital CCD astronomical cameras in the 1990s it became possible to obtain higher resolution images of the planets than had previously been possible through photographic means which used either color or black and white films. Essentially though these new cameras had the same Achilles heel as photographic exposures, being that it was not possible to be certain at the time of exposure, seeing was steady enough to maintain the original sharpest focus. It was quickly discovered that as a prerequisite a motorized focuser was essential to avoid having to touch the telescope which made focusing much more difficult. Instead the best focus could be achieved by watching the laptop/PC that the camera was coupled to. Even with this modified focuser it was still very difficult to obtain the best focus because of the very slow refresh rate of the camera of only a few frames per second. Thus if you were fortunate you could perhaps obtain one image in 20 or so that was sharper than the rest but hardly ever critically sharp.

Faced with this very hit and miss situation by the early 2000s amateurs started experimenting with webcams which were designed for home use to enable communications between individuals using the Internet. Most of these cameras were

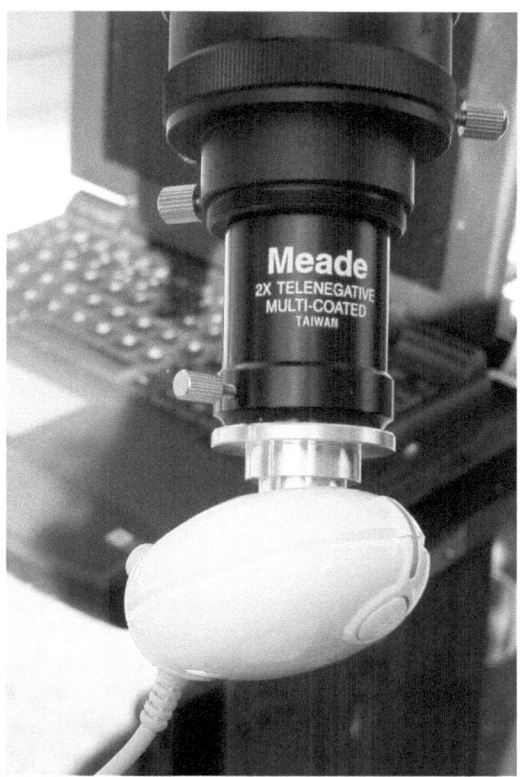

A Philips PCVC740 webcam set up with a Barlow lens at prime focus for lunar and planetary imaging (Credit: Grego)

fitted with a CMOS sensor with a 640 × 480 pixel array made up of 5.6 μm pixels which were fine for their intended purpose but for planetary imaging they were not sensitive enough. The one notable exception was the Philips ToUcam range of cameras which were equipped with a more sensitive Sony ICX98 CCD color sensor, initially with a USB1 connection, but later models were able to use a USB2 connection. These cameras had frame rates of between 5 and 30 frames per second and on removing the supplied lens could have an adaptor to fit the telescope draw tube fitted. The ability to be able to focus the camera in real time, because of the much faster refresh rate, was a huge improvement over the cooled CCD cameras. The USB1 cameras had a very limited bandwidth available and as a result the downloaded frames were heavily compressed and were only useable for planetary imaging at either 5 or 10 frames per second – beyond that the compression was too severe unless many thousands of frames were captured. The USB2 cameras had a greater bandwidth available but still were compressed.

The use of these cameras was only made possible by the development of Freeware, notably Registax, to align, stack and sharpen up the best frames which were captured either as an avi or bmp formats – this combination of many frames produced a final image that with good seeing were remarkable in the planetary detail shown. An added advantage was the relatively low cost of the ToUcams which gave amateurs the chance to see if they suited their systems and if not there was no great loss of capital.

Those amateurs who had become wedded to the use of the ToUcam cameras were only too well aware of their limitations and were constantly on the lookout for something better. By 2004 it became apparent to them that these specialized cameras were exactly what were required as they offered frames rates of up to 60 frames per second using either Firewire 400 or USB2 connectivity with no compression whatsoever. They were available as either color or black and white cameras and, for example, The Imaging Source DMK 21 range of cameras had the same Sony CCD ICX98 sensor as fitted to the ToUcam, Lumenera cameras had a Sony 640×480 pixel sensor but with 7.4 μm pixels, with both being much more expensive than the ToUcam but it was felt that the benefits they offered were far in excess of the additional cost. The use of a black and white camera, rather than a color version, gave the ability to image in monochrome as well as by using RGB color filters and combining them to produce a higher resolution at the same focal length than a color camera. This is because instead of the four pixels required to produce color images from a color camera, the color had a resolution of a single pixel in a mono camera albeit it with considerably more processing work being involved.

The latest developments with these cameras are the use Of the new Sony ICX618 sensor which has twice the sensitivity of the ICX098 sensor in the blue part of the spectrum, 2.5× in the green rising to 3× in the red and 4× in the near infrared. The Imaging Source produces a USB2 camera capable of 60 frames per second, while Basler Ace have a 100 frames per second camera using a GigE connection and Point Grey Flea 3 using a Firewire 800 connection again at 100 frames per second, all without any compression. As one would expect these camera are again more expensive than the 'entry level' machine vision versions mentioned in the paragraph above.

Only The Imaging Source cameras come with software for operating the camera and saving the captured frames but there is a freeware program, FireCapture, which does the job for the Basler and Point Grey cameras. Who knows what future developments are in the offing, even more sensitive sensors, USB3 will be a candidate but what would be most welcome is something on the seeing which is regarded at this point in time as being virtually impossible.

9.22   Observing the Martian Moons

Phobos and Deimos are exceedingly faint objects. To observe visually they require at least a 300 mm telescope, favorable Martian opposition circumstances, a steady atmosphere and a keen eye to observe visually.

It is important to be aware of the precise positions of the satellites at the time of the observation – a good astronomical computer program or a website such as the NASA/JPL Solar System Simulator (URL in appendix) can be used to find out this information. Both moons move around Mars at such a rapid speed that there is only a small window of opportunity to view them at their maximum elongations from the planet, hence as far away as possible from the red planet's glare. Deimos is the least difficult to discern when at a maximum elongation from Mars – around

3.5 Martian diameters from the planet's center and shining a feeble magnitude 12.9 at a close perihelic opposition.

If Mars is fully visible in the field of view of an ordinary eyepiece its glare renders the satellites virtually impossible to see; therefore the disk of Mars needs to be hidden from direct view, eliminating most of the intrusive light. Some residual glare will always be visible due to diffraction and irradiation, even if Mars is just out of the field of view. Specialized occulting eyepieces are available commercially, but it is within the practical amateur's means to fabricate a makeshift occulting mask from a small piece of tin foil inserted near the eyepiece's focal plane; this will enable Mars to be positioned within the field at the same time that the satellite search is taking place.

Chapter 10

Observer's Guide to Martian Apparitions 2012–2022

10.1 Seasonal Weather Phenomena

10.1.1 Northern Spring, Southern Autumn

Early on in the season the extensive north polar hood begins to dissolve while the south polar cap starts its retreat. Emerging into the warmth of the Sun a once frosty Hellas clears and darkens. Orographic cloud may be visible over Apollinaris Mons and there is the possibility of dust clouds in the southern hemisphere. Mid-season the north polar cap is eventually exposed, with the possibility of the appearance of Rima Tenuis. In the last part of the season, the limb regions often appear bright; there is increasing cloud in the north and hazes may be seen in both the north and south.

10.1.2 Northern Summer, Southern Winter

Orographic clouds appear over the Tharsis region and Nix Olympica. The north polar cap may remain static or continue to shrink further. Dust clouds and dust storms may whip up at any time. Note the darkness and definition of Mare Acidalium. Towards the end of the season clouds and frosts are likely to occur in the north.

10.1.3 Northern Autumn, Southern Spring

Initially the south polar cap is at its largest extent, but as the Sun climbs higher in the south it soon begins to retreat rapidly. Dust clouds may arise over Hellespontus, and late on in the season there's always the possibility of large scale dust storms, even global ones. Orographic clouds often appear over Elysium and Arsia Mons.

P. Grego, *Mars and How to Observe It*, Astronomers' Observing Guides,
DOI 10.1007/978-1-4614-2302-7_10, © Springer Science+Business Media New York 2012

10.1.4 Northern Winter, Southern Summer

Check on the width and intensity of Syrtis Major – it often appears darken and narrow during this season. The north polar hood is visible, often extending to northern temperate latitudes; when it clears, a projection called Novissima Thyle is sometimes observed. In the retreating south polar cap Rima Australis becomes visible, along with a projection called Argenteus Mons. Dust clouds whipped up from the south may encroach into Hellespontus, Hellas and Noachis. Mid-season, orographic cloud may be seen over Arsia Mons. Later in the season the north polar hood is extensive.

10.2 2011–2013 Apparition

Apparition begins (conjunction): 4 February 2011
Apparition ends (conjunction): 17 April 2013

10.2.1 Observing Season (Mars Greater Than 5 Arcseconds Across)

Observing season begins: 19 September 2011
Position: RA 08 h 09 m, Dec 21° 06′ (western Gemini)
Observing season ends: 19 September 2012
Position: RA 14 h 59 m, Dec −17° 42′ (western Libra)

10.2.2 Opposition Details

Opposition: 3 March 2012
Diameter: 13.9″
Martian season: Northern summer, southern winter
Opposition distance from Earth: 101.1 million kilometers
Magnitude: −1.1
Position: RA 11 h 08 m, Dec 10° 10′ (eastern Leo)
Opposition CM (00 h UT): 254°
Position angle of north pole: 18.5°
Tilt: 22.2°

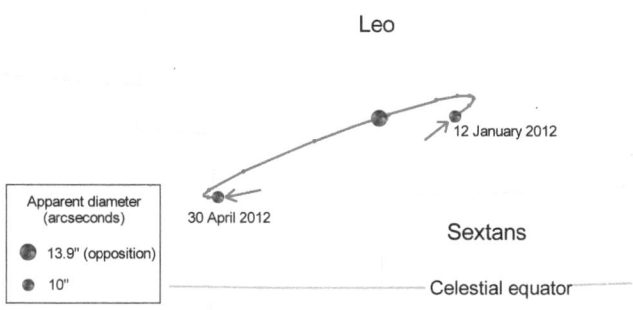

The celestial path of Mars at its best this apparition (when it is greater than 10 arcseconds across) between 12 January 2012 and 30 April 2012. Planets mark significant size milestones and dots are spaced at weekly intervals (Credit: Grego)

10.2.3 Distance and Diameter

8 March 2011: Perihelion (206.6 million kilometers from Sun).

12 January 2012: Diameter exceeds 10″.

14 February 2012: Aphelion (249.2 million kilometers from Sun).

25 January 2012: Mars retrogrades.

5 March 2012: Closest approach to Earth (100.8 million kilometers, diameter 13.9″)

16 April 2012: Mars prograides.

30 April 2012: Diameter falls below 10″.

5 July 2012: Mars moves south of the celestial equator.

24 January 2013: Perihelion (206.6 million kilometers from Sun).

15 March 2013: Mars moves north of the celestial equator.

10.2.4 Seasonal Phenomena

14 September 2011: Martian northern autumn, southern spring equinox.

30 March 2012: Martian northern summer, southern winter solstice.

30 September 2012: Martian northern autumn, southern spring equinox.

24 February 2013: Martian northern winter, southern summer solstice.

10.2.5 Appulses, Conjunctions and Occultations

30 September to 2 October 2011: Mars (mag +1.3) passes across the bright open cluster Praesepe (M44, mag +3.7).

10 November 2011: Mars (mag +1.0) is 1.5° north of Regulus (Alpha Leonis, mag +1.3).

17 March 2012: Mars (mag −1.1) 1° south of galaxy NGC 3384 (mag +10).

17 August 2012: Mars (mag +1.1) 3.8° south of Saturn (mag +0.8).

19 September 2012: Moon (4 days old) occults Mars (mag +1.2); daytime East Pacific, early evening South America and SW Atlantic).

10.3   2013–2015 Apparition

Apparition begins (conjunction): 17 April 2013
Apparition ends (conjunction): 14 June 2015

10.3.1   Observing Season (Mars Greater Than 5 Arcseconds Across)

Observing season begins: 6 November 2013
Position: RA 10 h 58 m, Dec 08° 22′ (southern Leo)
Observing season ends: 10 December 2014
Position: RA 12 h 06 m, Dec 01° 25′ (western Virgo)

10.3.2   Opposition Details

Opposition: 8 April 2014
Diameter: 15.1″
Martian season: Northern summer, southern winter
Opposition distance from Earth: 93.1 million kilometers
Magnitude: −1.5
Position: RA 13 h 14 m, Dec −05° 09′ (Virgo)
Opposition CM (00 h UT): 79°
Position angle of north pole: 34.5°
Tilt: 21.4°

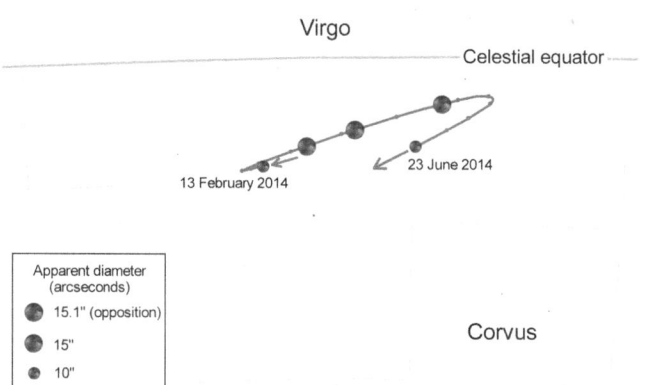

Serpens Caput

Virgo

Celestial equator

13 February 2014

23 June 2014

Apparent diameter
(arcseconds)

15.1" (opposition)

15"

10"

Corvus

The celestial path of Mars at its best this apparition (when it is greater than 10 arcseconds across) between 13 February 2014 and 23 June 2014. Planets mark significant size milestones and dots are spaced at weekly intervals (Credit: Grego)

10.3.3 Distance and Diameter

16 December 2013: Mars moves south of the celestial equator.

1 January 2014: Aphelion (249.2 million kilometers from Sun).

13 February 2014: Diameter exceeds 10″.

2 March 2014: Mars retrogrades.

6 April 2014: Diameter exceeds 15″.

14 April 2014: Closest approach to Earth (92.4 million kilometers, diameter 15.2″).

22 April 2014: Diameter falls below 15″.

21 May 2014: Mars progrades.

23 June 2014: Diameter falls below 10″.

11 December 2014: Perihelion (206.6 million kilometers from Sun).

21 February 2015: Mars moves north of the celestial equator.

10.3.4 Seasonal Phenomena

1 August 2013: Martian northern spring, southern autumn equinox.

15 February 2014: Martian northern summer, southern winter solstice.

23 June: Martian northern mid-summer, southern mid-winter.

18 August 2014: Martian northern autumn, southern spring equinox.

12 January 2015: Martian northern winter, southern summer solstice.

10.3.5 Appulses, Conjunctions and Occultations

9 May 2013: Moon (29.2 days old) occults Mars (mag +1.3); USA, North Atlantic, UK and Western Europe (day).

24 December 2013: Mars (mag +1.0) 1.3° south of galaxy Reinmuth 80 (mag +10.5).

29 December 2013: Mars (mag +0.9) 0.6° south of Porrima (Gamma Virginis, mag +3.4)

5 May 2014: Mars (mag −1.1) 1.4° south of Porrima.

6 July 2014: Moon (7.9 days old) occults Mars (mag +0.3); late afternoon East Pacific, early evening central America, evening South America.

13 July 2014: Mars (mag +0.2) 1.5° north of Spica (Alpha Virginis, mag +1.0).

22 August 2014: Mars (mag +0.6) 1.5° south of Zubenelgenubi (Alpha Librae, mag +2.8).

27 August 2014: Mars (mag +0.2) 3.5° south of Saturn (mag +0.6).

18 September 2014: Mars (mag +0.7) 0.5° north of Dschubba (Delta Scorpii, mag +2.3).

17 October 2014: Mars (mag +0.9) 1° south of planetary nebula NGC 6369 (mag +13.0).

4 November 2014: Mars (mag +0.9) 0.6° north of Kaus Borealis (Lambda Sagittarii, mag +2.8).

12 November 2014: Mars (mag +0.9) 2° north of Nunki (Sigma Sagittarii, mag +2.0).

21 February 2015: Mars (mag +1.3) just 0.5° north of Venus (mag −4.0).

11 March 2015: Mars (mag +1.5) just 0.25° north of Uranus (mag +5.9).

21 March 2015: Moon (1.8 days old) occults Mars (mag +1.3); daytime South Pacific and Antarctica.

2015 February 20. Evening (N hem favoured). Mars (top, mag +1.3) and Venus (mag −4.0) are separated by a degree, near the waxing crescent Moon (Credit: Grego)

10.4 2015–2017 Apparition

Apparition begins (conjunction): 14 June 2015
Apparition ends (conjunction): 26 July 2017

10.4.1 Observing Season (Mars Greater Than 5 Arcseconds Across)

Observing season begins: 10 December 2015
Position: RA 13 h 02 m, Dec −04° 57′ (mid-Virgo)
Observing season ends: 7 February 2017
Position: RA 00 h 27 m, Dec 02° 38′ (southern Pisces)

10.4.2 Opposition Details

Opposition: 22 May 2016
Diameter: 18.4″
Martian season: Martian late northern summer, late southern winter.
Opposition distance from Earth: 76.3 million kilometers
Magnitude: −2.1
Position: RA 15 h 58 m, Dec −21° 40′ (northwestern Scorpius)
Opposition CM (00 h UT): 178°
Position angle of north pole: 36.8°
Tilt: 10.4°

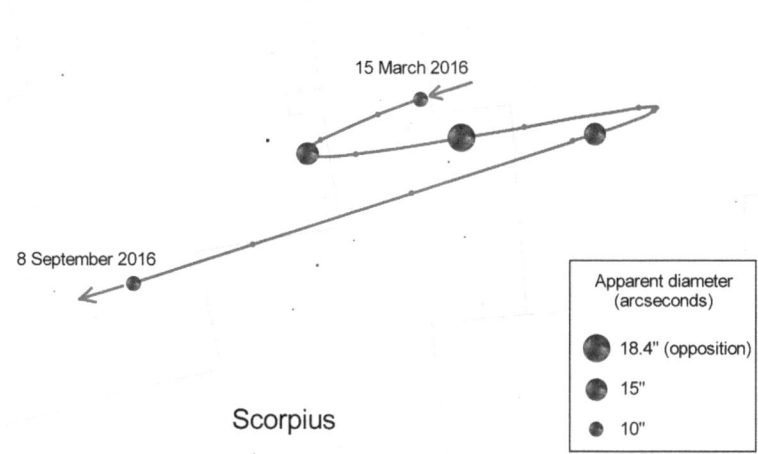

Ophiuchus

Libra

15 March 2016

8 September 2016

Scorpius

Apparent diameter
(arcseconds)

18.4" (opposition)

15"

10"

The celestial path of Mars at its best this apparition (when it is greater than 10 arcseconds across) between 15 March 2016 and 8 September 2016. Planets mark significant size milestones and dots are spaced at weekly intervals (Credit: Grego)

10.4.3 Distance and Diameter

18 November 2015: Mars moves south of the celestial equator.

19 November 2015: Aphelion (249.2 million kilometers from Sun).

15 March 2016: Diameter exceeds 10″.

16 April 2016: Mars retrogrades.

20 April 2016: Diameter exceeds 15″.

30 May 2016: Closest approach to Earth (75.3 million kilometers, diameter 18.6″).

1 July 2016: Mars progrades.

16 July 2016: Diameter falls below 15″.

8 September 2016: Diameter falls below 10″.

29 October 2016: Perihelion (206.6 million kilometers from Sun).

29 January 2017: Mars moves north of the celestial equator.

10.4.4 Seasonal Phenomena

19 June 2015: Martian northern spring, southern autumn equinox.

3 January 2016: Martian northern summer, southern winter solstice.

5 July 2016: Martian northern autumn, southern spring equinox.

29 November 2016: Martian northern winter, southern summer solstice.

6 May 2017: Martian northern spring, southern autumn equinox.

10.4.5 Appulses, Conjunctions and Occultations

16 July 2015: Mars (mag +1.6) 0.1° north of Mercury (mag −1.5).

20–21 August 2015: Mars (mag +1.8) passes across the bright open cluster Praesepe (M44, mag +3.7).

25 September 2015: Mars (mag +1.8) 0.8° north of Regulus (Alpha Leonis, mag +1.3).

18 October 2015: Mars (mag +1.8) 0.5° northeast of Jupiter (mag −1.8).

3 November 2015: Mars (mag +1.7) 0.5° north of Venus (mag −4.3).

27 November 2015: Mars (mag +1.6) 2° south of galaxy Reinmuth 80.

6 December 2015: Moon (24.7 days old) occults Mars (mag +1.5); northeast Africa (morning), Western Australia and Indian Ocean (daytime).

1 February 2016: Mars (mag +0.8) 1.2° north of Zubenelgenubi.

16 March 2016: Mars (mag −0.1) 0.1° north of Graffias (Beta Scorpii, mag +2.5).

20 May 2016: Mars (mag −2.0) 0.5° north of Dschubba.

10 August 2016: Mars (mag −0.6) 0.8° south of Dschubba.

24 August 2016: Mars (mag −0.4) 1.6° north of Antares (Alpha Scorpii, mag +1.0) and 4.4° south of Saturn (mag +0.4).

4 September 2016: Mars (mag −0.1) 0.8° south of Theta Ophiuchi (mag +3.3).

29 September 2016: Mars (mag 0.0) 2° north of open cluster NGC 6520 (mag +8.0).

7 October 2016: Mars (mag +0.1) 0.3° south of Kaus Borealis.

16 October 2016: Mars (mag +0.2) 1.1° north of Nunki.

10 December 2016: Mars (mag +0.7) 1.5° north of Nashira (Gamma Capricorni, mag +3.7) and northwest of Deneb Algiedi (Delta Capricorni, mag +2.8).

31 December 2017: Mars (mag +0.9) 0.3° west of Neptune (mag +7.9).

3 January 2017: Moon (4.7 days old) occults Mars (mag +1.0); Philippines (late afternoon), South China Sea and northwest Pacific (early evening).

1 February 2017: Mars (mag +1.1) 8.8° west of the Moon (4.4 days old) and 5.5° east of Venus (mag −4.6).

27 February 2017: Mars (mag +1.3) 0.8° north of Uranus (mag +5.9).

2015 October 16. Morning view (from the northern hemisphere) of the eastern horizon showing a close grouping of Venus (*top*, mag −4.4), Mars (mag +1.8) and Jupiter (mag −1.8) and Mercury (*bottom*, mag −0.5). Inset: A 1° degree field showing Mars, with Jupiter and three of its bright moons (Credit: Grego)

10.5   2017–2019 Apparition

Apparition begins (conjunction): 26 July 2017
Apparition ends (conjunction): 2 September 2019

10.5.1   Observing Season (Mars Greater Than 5 Arcseconds Across)

Observing season begins: 8 January 2018
Position: RA 15 h 07 m, Dec −16° 40′ (central Libra)

Observing season ends: 15 March 2019
Position: RA 03 h 08 m, Dec 18° 26′ (eastern Aries)

10.5.2   Opposition Details

Opposition: 27 July 2018
Diameter: 24.2″
Martian season: Martian northern autumn, southern spring equinox
Opposition distance from Earth: 58.1 million kilometers

Magnitude: −2.8

Position: RA 20 h 33 m, Dec −25° 28′ (southwestern Capricornus)

Opposition CM (00 h UT): 78°

Position angle of north pole: 5.7°

Tilt: −11.2°

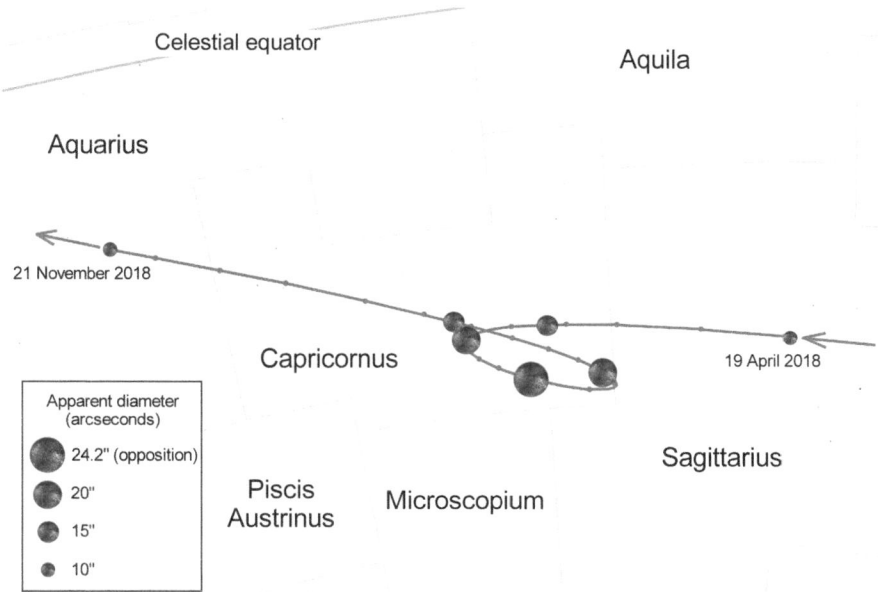

The celestial path of Mars at its best this apparition (when it is greater than 10 arcseconds across) between 19 April 2018 and 21 November 2018. Planets mark significant size milestones and dots are spaced at weekly intervals (Credit: Grego)

10.5.3 Distance and Diameter

7 October 2017: Aphelion (249.2 million kilometers from Sun).

27 October 2017: Mars moves south of the celestial equator.

19 April 2018: Diameter exceeds 10″.

27 May 2018: Diameter exceeds 15″.

23 June 2018: Diameter exceeds 20″.

29 June 2018: Mars retrogrades.

31 July 2018: Closest approach to Earth (57.6 million kilometers, diameter 24.3″).

28 August 2018: Mars progrades.

8 September 2018: Diameter falls below 20″.

16 September 2018: Perihelion (206.6 million kilometers from Sun).

9 October 2018: Diameter falls below 15″.

21 November 2018: Diameter falls below 10″.

2 January 2019: Mars moves north of the celestial equator.

26 August 2019: Aphelion (249.2 million kilometers from Sun).

10.5.4   Seasonal Phenomena

20 November 2017: Martian northern summer, southern winter solstice.

23 May 2018: Martian northern autumn, southern spring equinox.

17 October 2018: Martian northern winter, southern summer solstice.

24 March 2019: Martian northern spring, southern autumn equinox.

10.5.5   Appulses, Conjunctions and Occultations

2 September 2017: Mars (mag +1.8) 4° north of Mercury (mag +2.6).

17 September 2018: Mars (mag +1.8) 0.4° west of Mercury (mag −0.8).

18 September 2017: Moon (28 days old) occults Mars (mag +1.8); eastern Pacific (morning).

6 October 2017: Mars (mag +1.8) 0.2° west of Venus (mag −3.9).

9 November 2017: Mars (mag +1.8) 1.7° south of Porrima.

20 November 2017: Martian northern summer, southern winter solstice.

28 November 2017: Mars (mag +1.7) 3.3° north of Spica.

7 January 2018: Mars (mag +1.4) 0.2° south of Jupiter (mag −1.8).

1 February 2018: Mars (mag +1.2) 0.3° south of Graffias.

9 February 2018: Mars (mag +1.1) 3.5° south of the Moon (23.6 days old).

6 March 2018: Mars (mag +0.7) 0.7° north of Little Ghost Nebula (NGC 6309, mag +13.0).

14 March 2018: Mars (mag +0.6) 3.3° south of planetary nebula NGC 6445 (mag +13.0).

2 April 2018: Mars (mag +0.3) 1.3° south of Saturn (mag +0.5).

2 November 2018: Mars (mag −0.6) 0.3° north of Nashira (Gamma Capricorni, mag +3.7).

5 November 2018: Mars (mag −0.5) 0.5° north of Deneb Algiedi.

16 November 2018: Moon (7.9 days old) occults Mars (mag −0.3); Antactica, South Pacific (early evening).

3 December 2018: Mars (mag 0.0) 0.9° south of Lambda Aquarii (mag +3.7).

7 December 2018: Mars (mag +0.1) 0.1° north of Neptune (mag +7.9).

13 February 2019: Mars (mag +1.0) 1° north of Uranus (mag +5.8).

7 January 2018. Mars is very close to Jupiter and its moons. This view simulates the appearance of the scene at 05 h UT using a 15 mm Plossl on a 200 mm SCT (133×) (Credit: Grego)

10.6   2019–2021 Apparition

Apparition begins (conjunction): 2 September 2019
Apparition ends (conjunction): 8 October 2021

10.6.1   Observing Season (Mars Greater Than 5 Arcseconds Across)

Observing season begins: 8 February 2020
Position: RA 17 h 35 m, Dec −23° 24′ (southern Ophiuchus)
Observing season ends: 15 April 2021
Position: RA 05 h 38 m, Dec 24° 49′ (northeastern Taurus)

10.6.2   Opposition Details

Opposition: 13 October 2020
Diameter: 22.4
Martian season:
Opposition distance from Earth: 63.1 million kilometers
Magnitude: −2.6
Position: RA 01 h 24 m, Dec 05° 29′ (southeastern Pisces)

Opposition CM (00 h UT): 181°
Position angle of north pole: 325.0°
Tilt: −20.3°

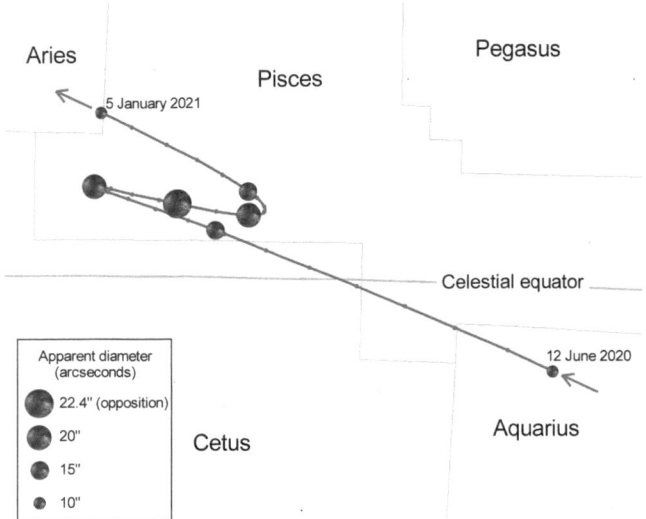

The celestial path of Mars at its best this apparition (when it is greater than 10 arcseconds across) between 12 June 2020 and 5 January 2021. Planets mark significant size milestones and dots are spaced at weekly intervals (Credit: Grego)

10.6.3 Distance and Diameter

8 October 2019: Mars moves south of the celestial equator.

8 February 2020: Diameter exceeds 5″.

12 June 2020: Diameter exceeds 10″.

12 July 2020: Mars moves north of the celestial equator.

1 August 2020: Diameter exceeds 15″.

2 August 2020: Perihelion (206.5 million kilometers from Sun).

5 September 2020: Diameter exceeds 20″.

10 September 2020: Mars retrogrades.

6 October 2020: Closest approach to Earth (62.1 million kilometers).

4 November 2020: Diameter falls below 20″.

16 November 2020: Mars progrades.

1 December 2020: Diameter falls below 15″.

5 January 2021: Diameter falls below 10″.

17 September 2021: Mars moves south of the celestial equator.

10.6.4 Seasonal Phenomena

8 October 2019: Martian northern summer, southern winter solstice.

3 September 2020: Martian northern winter, southern summer solstice.

8 February 2021: Martian northern spring, southern autumn equinox.

10.6.5 Appulses, Conjunctions and Occultations

12 December 2019: Mars (mag +1.7) 0.3° east of Zubenelgenubi.

8 January 2020: Mars (mag +1.5) 0.7° south of Graffias.

6 February 2020: Mars (mag +1.3) 0.5° north of Little Ghost Nebula.

26 February 2020: Mars (mag +1.1) 1.7° north of Kaus Borealis.

18 March 2020: Lovely grouping of Mars (mag +0.9), Jupiter (mag −2.1) and the Moon (23.9 days old).

20 March 2020: Mars (mag +0.9) 0.7° south of Jupiter (mag −2.1).

23 March 2020: Mars (mag +0.9) less than 1 arcminute south of Pluto (mag +14.4). If you've never managed to see Pluto, now's the ideal opportunity, courtesy of Mars; its two moons, Phobos (mag +13.7) and Deimos (mag +14.8) are also on view at good elongations. Using a low power eyepiece you can fit Jupiter (mag −2.1), 1.7° further west, into the same low power field of view.

26 March 2020: Mars (mag +0.9) directly between Jupiter (mag −2.1) and Saturn (mag +0.7), 3.5° from each.

31 March 2020: Mars (mag +0.8) 1° south of Saturn (mag +0.7).

5 May 2020: Mars (mag +0.4) 1° north of Deneb Algiedi.

12 June 2020: Mars (mag −0.2) 1.7° south of Neptune (mag +7.9).

21 October 2019: Mars (mag +1.9) 2° south of Porrima.

9 November 2019: Mars (mag +1.8) 3° north of Spica.

21 January 2021: Mars (mag +0.2) 1.8° north of Uranus (mag +5.8).

3 March 2021: Mars (mag +0.9) 2° south of the bright Pleiades open star cluster (M45).

9 May 2021: Mars (mag +1.6) 0.8° south of Mebsuta (Epsilon Geminorum, mag +3.0).

1 June 2021: Mars (mag +1.7) 1.8° south of Kappa Geminorum (mag +3.6).

22–24 June 2021: Mars (mag +1.8) crosses Praesepe.

13 July 2021: Mars (mag +1.8) 0.5° south of Venus (mag −3.9).

29 July 2021: Mars (mag +1.8) 0.6° north of Regulus.

23 March 2020. Pluto and Mars are exceedingly close to each other, presenting a visual challenge and a great astroimaging opportunity. This is the view at 05 h UT using a high magnification (Credit: Grego)

3 March 2021. Mars roves south of the Pleiades. This is the view at 21 h UT using 15×70 binoculars (Credit: Grego)

Chapter 11

The Mars Observer's Equipment

11.1 An Eye for Martian Detail

Regardless of whether a planetary observer owns a tiny reflector on a basic altazimuth mounting or a large expensive apochromatic refractor on a sturdy computer controlled mount, by far the most important optical equipment belonging to any planetary observer is a small but powerful pair of binoculars called eyes. Looking after these precious little instruments carefully will allow the owner to enjoy the beauty of the planets throughout their lifetime.

One aspect of the eye's structure that doesn't really affect planetary observing is the presence of a blind spot in each eye, caused by the lack of photoreceptive cells in the part of the retina where the optic nerve intrudes. Blind spots lie on the left side of the left eye's field of view, and on the right side of the right eye's field of view, but their presence isn't revealed when concentrating upon objects like Mars which have a small diameter through the telescope eyepiece. However, the following experiment demonstrates the complete blindness of the blind spot, and it usually astounds anyone first trying it. Cover your left eye with your left hand and, with your right eye, slowly scan the area to the left of Mars, a few degrees away from it. You will eventually be able to find a spot to look at where Mars has completely vanished from sight. The actual area covered by the blind spot measures about 6° across – ten times the Moon's apparent diameter when seen with the naked eye.

Many people suffer from floaters – minute dark flecks, translucent cobwebs or clouds of various shapes and sizes that occasionally become visible when viewing a bright object like Mars. Floaters are the shadows cast onto the retina by the remnants of dead cells floating in the vitreous humor. Everyone has floaters, and they can be annoying because they can obscure small features. Most people put up with them. Floaters increase in number with age, and laser eye surgery or a vitrectomy can remove them if the condition causes severe disruption of normal vision.

Annual eye checkups are recommended, as some unsuspected but treatable medical conditions may come to light as a result. Smoking tobacco is detrimental to astronomical observation, let alone the smoker's general health. Eyes use nearly all of the oxygen that reaches them through the blood vessels; the carbon monoxide in cigarette smoke attaches to the hemoglobin in red blood cells more readily than oxygen itself. Drinking alcohol while observing decreases how much detail can be seen on Mars in direct proportion to the amount of alcohol con-

P. Grego, *Mars and How to Observe It*, Astronomers' Observing Guides, DOI 10.1007/978-1-4614-2302-7_11, © Springer Science+Business Media New York 2012

sumed, and the observer's efficiency is totally compromised when, despite the observer's best efforts, the eye cannot be made to remain in proximity to the eyepiece. Alcohol dilates the blood vessels, and though it may make the consumer feel warm for a short while, the extra heat loss from the body can be dangerous on cold nights. So, alcohol ought to be avoided until after the observing session when one is safely indoors at home. Finally, visual acuity actually diminishes if blood sugar levels are low, so a small snack during the observing session is both pleasant and beneficial.

11.2   Binoculars

Binoculars are somewhat underrated by amateur astronomers, and few would seriously consider using them to regularly observe the planets, especially Mars. Yet binoculars have certain advantages over telescopes. They are usually far less expensive than telescopes, they are easier to carry around (even with a stand), they deliver a wide field of view and they are more robust than telescopes, capable of withstanding occasional knocks and still remain in optical collimation.

It is important to hold binoculars as steadily as possible, either propped up firmly against a solid object (like a car roof) or better still, fixed to a tripod or a dedicated binocular mount. A steady field of view adds enormously to the observer's pleasure – those restricted to hand-held viewing may view the skies for a few minutes at a time, but observers who are able to hold their binoculars steadily are likely to observe for far longer periods of time. When Mars is steadily positioned in the field of view, a quality pair of binoculars enhances the planet's color and reveals the planet's immediate celestial surroundings in great detail.

The power of binoculars is identified by two figures that denote their magnification and the size of their objective lenses – 7×30 binoculars give a magnification of 7× and have 30 mm objective lenses. Small to medium sized binoculars (with objective lenses from 25 to 50 mm in diameter) usually deliver low magnifications that range between powers of 7 and 15 times. Since most binoculars have a relatively low magnification and Mars is such a small object in terms of apparent diameter, the planet's disk, if resolved at all, shows little in the way of detail. However, even at aphelic opposition, when the planet is around 14 arcseconds across, specialized big binoculars with higher magnifications (25×100 s at least) can show some detail, including the planet's phase, some of the larger, darker markings and the bright polar caps.

The true field of view (the actual area of sky) that is discerned through binoculars deceases with magnification. My own 7×50 binoculars (a budget brand, but of good optical quality) deliver 7× magnification and a true field of view some 7° wide. My 15×70 binoculars have a true field of view of 4.4°, some nine times the diameter of the Moon. Binoculars equipped with wide angle eyepieces (usually high-end instruments) produce bigger actual fields of view and their apparent fields of view are greater than 60°. With their wide true fields of view, binoculars take in a large area of the sky and stunningly beautiful views can be had when Mars is in the vicinity of bright star groupings and the occasional planet.

Large binoculars (those with 60 mm objectives, and larger) with high magnifications provide more detailed views, but they must be held steadily by some means. In fact, any binoculars giving more than 10× magnification must be supported firmly, however lightweight they may be, because higher magnifications exacerbate any slight movement of the observer's body. There is an exception – image stabilized binoculars which eliminate minor shakes of the user's body. At first glance they resemble regular binoculars of the same aperture but weigh slightly more. At the push of a button they deliver sharp, vibration-free views courtesy of moving optical elements (most of them require batteries). Stabilized binoculars typically offer quite high magnifications (up to 18×), with apertures from 30 to 50 mm.

Some amateur astronomers are tempted to seek the best of both worlds by having the convenience of binoculars with a zoom facility. A typical pair of zoom binoculars can be adjusted to magnify from, say 15× to as much as 100×. On the face of it, such an instrument would be ideal, but there are drawbacks. When set at a low magnification, the apparent field of view in zoom binoculars is usually unimpressive – perhaps as small as 40°, which is far smaller than the apparent field visible in a pair of regular binoculars of comparable magnification. Zooming involves physically altering the distance between the lenses inside the eyepiece using an external lever – this is not usually such a simple operation, as some refocusing is usually necessary after a change in magnification. Crucially, the alignment of the left and right optical systems in any zoom binoculars needs to be absolutely spot-on in order that the brain can produce a merged image from two separate high power images, and this is where most budget zoon binoculars prove wanting. Even if a pair of binoculars does provide good high magnification views, the rotation of the Earth makes celestial objects appear to move across the field of view rather rapidly. For example, at a magnification of 50×, an apparent field of 50° will equate to a true field of view of 1°, and Mars will move from one edge of the field to the other in just 2 min. This means frequent adjustments to the mount if the planet is to be kept under scrutiny for any length of time.

Numerous different optical configurations are used in binoculars, and this is obvious when you look at the wide variety of different shapes and sizes available. As is usually the case, you get what you pay for, but providing they are purchased from a reputable optical dealer, the quality of budget optical goods these days is usually quite good. The quality of the optical system, the optical materials used and the build of binoculars is however noticeable when budget binoculars are compared with high-end binoculars in actual use. High-end binoculars use the best optical glass in their object glasses, internal prisms and eyepiece lenses, and they are figured and aligned to exacting standards. The optical surfaces are usually multi-coated to minimise reflections, and internal baffles block stray light and internal reflections, providing better contrast.

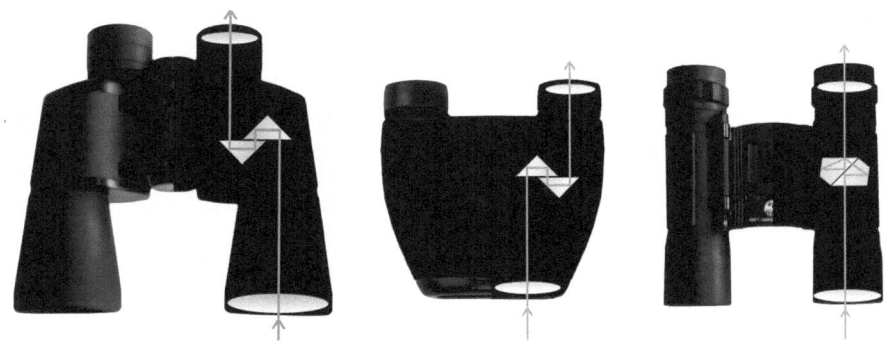

Porro prism, inverted Porro prism and roof prism binoculars, showing their internal folding light paths (Credit: Grego)

Most binoculars use glass prisms that fold the light between the objective lens and eyepiece lens, and they produce a right way up image, which is of course essential for everyday terrestrial use. 7 × 50 binoculars are ideal general purpose astronomy binoculars. They deliver a wide field of view and have a magnification low enough for the observer to peruse the skies for short periods without requiring the use of a binocular support. 7 × 50 binoculars have an exit pupil of 7 mm. The exit pupil is the diameter of the circle of light that projects from an eyepiece into the eye, and its size can be derived from the aperture of the binoculars divided by their magnification. As the dark adapted eye has an average size of 7 mm, the optimum size of exit pupil for deep dark sky astronomical use measures 7 mm.

There are two basic types of prism binoculars – Porro prism binoculars and Roof prism binoculars. Until a couple of decades ago, most binoculars were of the Porro prism type. Most often, Porro prism binoculars have a distinct 'W' shape, produced by the way that the prisms are arranged to fold the light from the widely separated objective lenses to the eyepieces. American-style Porro prism binoculars are a sturdy design, with prisms mounted on a shelf inside a single molded case. German-style Porro prism binoculars feature objective housings that screw into the main body containing the prisms – their modular nature makes them more susceptible to going out of collimation after knocks. In recent years, a new style of small binoculars has appeared that use an inverted Porro prism design, leading to a 'U' shaped shell with objective lenses that may be closer together than the eye lenses. The majority of small binoculars produced today are of the Roof prism design, popular because they are compact and lightweight. Roof prism binoculars often have a distinctive H-shape that looks like two small telescopes place side by side – a casual observer may think this indicates a straight-through optical configuration without any intervening prisms. Because of the way that Roof prism binoculars fold the light, they offer generally less contrasting views than Porro prism binoculars.

11.3　Telescopes

Every amateur astronomer who has owned a telescope will have eventually turned it towards Mars – few have ever been disappointed at the view, particularly around the time of a favourable opposition.

Choosing the right telescope can be a difficult task for a novice. What may be a good telescope for low power deep sky observation may not be the best one for observing the Moon and planets. An amazing array of telescopes of various kinds are advertised in astronomy magazines. Fortunately, these days the optical and build quality of most of these telescopes (an ever increasing proportion of them of them being Chinese imports) is quite acceptable for general astronomical observation and for viewing Mars at low to medium powers.

Regardless of a telescope's aperture or physical size, the most important thing to be aware of when purchasing any telescope is its optical quality. It isn't advisable to buy a new telescope from any source other than a reputable telescope dealer, and I recommend that sensational newspaper advertisements, department stores and general electrical stores ought to be avoided. Sadly, newspaper advertisements announcing massive stock clearances of telescopes and binoculars are often full of outrageous hyperbole, with claims of extravagant magnifications and the ability of the instrument to show all the wonders of the Universe. These ads attempt to blind the novice with 'science' in an attempt to hide the fact that their instruments may be made entirely of plastic – even the lenses! Such optical monstrosities are useless for any kind of observing, and the poor view that they deliver can be enough to put the novice off astronomy altogether! Telescopes sold through large shopping mall domestic goods retailers and department stores are usually overpriced. Moreover, large general retailers are not overly concerned about the optical quality of their goods, and the sales assistant will probably be unable to inform the buyer of the suitability of the telescope for astronomical purposes, other than stating the blurb printed on the box. The ultimate test of a retailer's confidence in the quality of their goods and a measure of their customer friendliness is to ask to examine and make a quick observational test of any telescope that you intend to buy. Any major flaws or in the optics or dings and defects on the exterior of the instrument are quickly revealed under the bright lights of the store.

Any reputable company that specializes in selling and/or manufacturing optical instruments will give by far the best deal, in terms of price, service and advice. All the major astronomical equipment retailers advertise in astronomy magazines, and most of them produce a product catalogue that can be browsed online or in printed form.

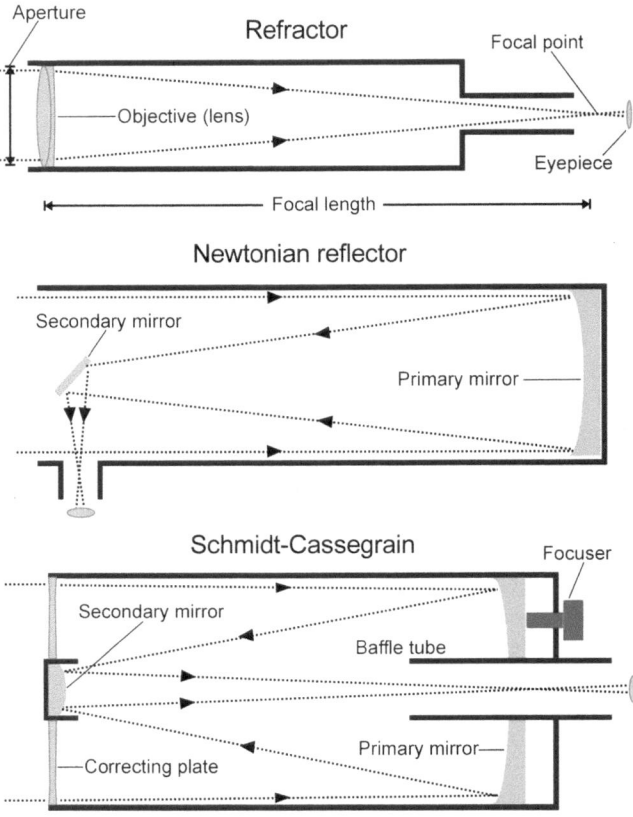

Refractor

Aperture

Objective (lens)

Focal point

Eyepiece

Focal length

Newtonian reflector

Secondary mirror

Primary mirror

Schmidt-Cassegrain

Focuser

Secondary mirror

Baffle tube

Primary mirror

Correcting plate

The basic optical configurations of an achromatic refractor, Newtonian reflector and Schmidt-Cassegrain telescope. The Maksutov-Cassegrain is similar to the Schmidt-Cassegrain, except that it uses a meniscus lens instead of a corrector plate and has an aluminised central spot inside the meniscus rather than an inbuilt secondary mirror structure (Credit: Grego)

11.4 Refractors

Asked to imagine a typical amateur astronomer's telescope, the refractor jumps into most people's mind. Refractors have an objective lens and an eyepiece at either end of a closed tube. Light is collected and focused by the objective lens (the light is refracted, hence the name of the optical system), and the eyepiece magnifies the focused image. Reference is often made to a telescope's focal length – this is the distance between the lens and the focal point, expressed as a multiple of the lens diameter or in millimeters. A 100 mm f/10 lens has a focal length of 1,000 mm. A 150 mm f/8 lens has a focal length of 1,200 mm. Eyepieces also have a focal length, but this is always expressed in millimeters, and never as a focal ratio – there is no such thing as an f/10 eyepiece, for example.

Galilean telescopes are the simplest form of refractor, with a single objective lens and a single eyepiece lens. They suffer greatly from chromatic aberration, caused by the splitting of light into its component colors after it is refracted through glass, and spherical aberration, caused by light not being brought to a single focus. Through a Galilean telescope Mars appears surrounded by fringes of vividly colored light, and the whole image appears washed out and blurry. Cheap small telescopes attempt to alleviate the worst effects of aberration by having large stops placed inside the telescope tube, preventing the outer parts of the cone of light from travelling down to the eyepiece. This crude trick simply makes a poor image appear slightly less poor, and the presence of the stop reduces the aperture of the instrument, reducing its light grasp and resolving power.

Small though they may be, good quality astronomical finderscopes and monoculars should not be confused with Galilean telescopes. Finderscopes are low power refractors that are attached to, and precisely aligned with larger telescopes, in order that the observer can locate celestial objects. When centered on the finder's cross hairs, the object is also visible at a higher magnification in the main instrument. Finderscopes have achromatic objective lenses (typically 20–50 mm) and have fixed eyepieces that can be adjusted for focus. Straight through finderscopes deliver an inverted view, so they are unsuitable for terrestrial use. Monoculars are little hand-held telescopes with small (typically 20–30 mm) achromatic objective lenses. Monoculars use roof prisms to deliver low power, right way up views. They can be carried in a coat pocket and they are great for cursory Mars observation. With their large true fields of view, bright stars and planets in the immediate vicinity of Mars can be seen.

A well-made telescope, of any size, is capable of providing a really pleasing view of Mars. Those who claim that any telescope with an aperture smaller than 75 mm is useless for observing Mars misunderstand the main reason why most people observe – for the sheer pleasure of seeing this distant world with their own eyes. Although a small telescope is incapable of revealing fine Martian detail, good oppositions provide enough detail to keep a novice enthralled.

On nights of poor seeing, when the atmosphere is shimmering and the stars scintillate wildly, attempts to observe Mars at high magnification through large instruments can prove futile, as it can take on the appearance of a shimmering, boiling orange blob, hardly worth bothering to look at. On these nights a small telescope will sometimes deliver an apparently sharper, more stable image than the large telescope, since a small telescope will not resolve as much atmospheric turbulence as a larger one. Small telescopes have a number of other advantages. Being lightweight, they are eminently portable, and they can be carried around an observing site to avoid local obstacles to the sky, such as trees and buildings. A small, relatively inexpensive telescope may be considered to be expendable, and for this reason the observer may actually be inclined to use it more often than a 'precious' high-end telescope – accidental damage to the external structure or the optics of a cheap telescope is not nearly as soul-destroying as bashing an instrument that cost ten times as much.

The least expensive small telescopes, including those hand-held old style brass 'naval' telescopes that use draw tubes to focus the image, have a fixed eyepiece that delivers a constant magnification. Some fixed eyepiece telescopes are a little more sophisticated by allowing some variation of the magnification delivered by the eyepiece. A telescope that allows the eyepieces to be interchanged, allowing the magnification to be alternated between low to high powers, is a somewhat more versatile instrument. Two or three eyepieces, and maybe a magnifying eyepiece called a Barlow lens, are often provided – these can be small, plastic mounted eyepieces with a 0.965-in. barrel diameter, and they are probably going to be of a very basic optical design and of poor quality. These eyepieces may deliver poor quality views with incredibly narrow apparent fields of 30° or even smaller.

Eyepieces are not accessories – they are as vital to the performance of a telescope as its objective lens or mirror. So, if a small instrument is not performing as well as might be expected, don't get rid of it straight away – replace the eyepieces with some better quality ones, purchased from an optical retailer. The most widely available eyepieces have a barrel diameter of 1.25-in., and these can be mounted to a .965-in. eyepiece tube using an adapter. Plössl eyepieces deliver an apparent field of about 50°, and quality budget versions of this eyepiece design are available. Good quality eyepieces can transform a small budget telescope into a good performer on Mars at low to medium magnifications (see below for more information about eyepieces).

Jacy Grego, the author's daughter, observes through a 'budget' 60 mm achromat. The telescope's original plastic eyepieces were replaced with quality ones; here a 0.965-in. Zeiss 16 mm orthoscopic is used (Credit: Grego)

Quality astronomical refractors have an achromatic objective lens that comprises two specially shaped lenses of different types of glass, nestled closely together. These lenses attempt to refract all the different wavelengths of light to a focus at a single point. Achromatic objectives do not eliminate chromatic aberration altogether, but generally speaking the effects are less noticeable in longer focal length refractors. Many budget imported achromatic refractors have focal lengths of f/8 to as short as f/5, and although they do display noticeable chromatic aberration, mainly in the form of a violet fringe around the Moon and the brighter planets, they offer good resolution and good contrast. One inexpensive way of reducing the false color is to use a minus-violet contrast boosting filter that screws into the eyepiece. Another, more expensive way to minimize chromatic aberration is to use a specially designed lens (one brand is called a 'Chromacorr') that attaches to the eyepiece, transforming a budget achromatic refractor into a telescope that approaches the performance of a high-end apochromat.

Apochromatic refractors use special glass in their two or three element objective lenses to bring the light to a tack-sharp focus, delivering images that are virtually free from the effects of chromatic aberration. Views of Mars through an apochromat are almost completely free of aberration and are of high contrast, rivaling the kind of view to be had through a good quality long focal length Newtonian reflector (see below). Aperture for aperture, an apochromatic refractor will be more than ten times as expensive than a budget achromatic refractor.

Refractors require little maintenance. Their objective lenses are aligned in the factory and sealed in a cell, and they are ready for immediate use straight out of the box. There is no reason to unscrew an objective lens and remove it from its cell, although being naturally curious, a great many amateur astronomers have been tempted to do it – just to see how the thing is put together! It is not recommended, as the reassembled lens will not perform as well, often noticeably so. Over time, the external surfaces of lenses can accumulate a fair amount of dust and debris, but great care must be taken when cleaning them. Most lenses have a thin layer of anti-reflection coating, which can be disrupted if the lens is cleaned improperly. Under no circumstances should a lens be rubbed vigorously with a cloth. Dust particles should be carefully removed with a soft optical brush or an air puffer, and any residual dirt can be gently removed with optical lens wipes, each used once and applied in a single stroke. Condensation on a lens should be allowed to dry naturally, and never rubbed off.

These impressive Victorian refractors (a 4- and 7-in. Cooke) housed in an observatory in Edgbaston, UK, were used by the author to observe Mars on many occasions during the 1980s (Credit: Grego)

11.5 Reflectors

Reflecting telescopes collect light with a specially shaped concave primary mirror and reflect it to a sharp focus. Reflected light is free from the effects of chromatic aberration, but they are prone to spherical aberration, more so in shorter focal length systems. The most popular reflecting telescope design, the Newtonian, uses a concave primary mirror held in a cell at the bottom of the tube and a smaller flat secondary mirror mounted on a 'spider' near the top of the tube that reflects light sideways through the tube into the eyepiece – the observer does not appear to be looking directly 'through' the telescope, but into its side, something that seems to confuse many laypersons! A well-collimated long focal length (f/10 and longer) Newtonian will deliver superbly detailed views of Mars at high magnification.

Cassegrain reflectors have a primary mirror with a central hole. The primary reflects light onto a small convex secondary mirror, which reflects the light back down the tube, through the hole in the primary mirror and into the eyepiece. Prone to the optical aberrations of astigmatism and field curvature, most Cassegrains are large observatory-sized instruments with focal lengths ranging from f/15 to f/25 – excellent for Martian studies at high magnification.

Reflectors require much more care and attention than refractors. Vibration or a slight sudden knock to the tube can cause misalignment of the primary mirror in its cell or the secondary in its spider. Misaligned optics will produce poor quality images, including dimming, blurring and multiple images near the point of focus. A brand new Newtonian taken out of its carton is likely to require recollimation in order to align the optical components as precisely as possible. The alignment of both the primary mirrors can usually be adjusted by hand using three wingnuts at the base of the mirror cell, but the secondary will usually require a small screwdriver or Allen key. Collimation can be time consuming and a little tricky for novices, but new telescopes should be provided with adequate instructions, and there are many Internet resources that explain the process in detail. There are ways to achieve good collimation, including laser collimators and a device called a Cheshire eyepiece. Cassegrains are more difficult to collimate than Newtonians. Most reflectors are not sealed when in use, and coated with a wafer thin layer of reflective aluminum, the primary and secondary mirrors gradually deteriorate by being exposed to the open air. Special coatings can extend the life of a mirror by a factor of two or three. However, all mirrors accumulate a layer of dust and bits of debris over time, and a primary mirror can look disconcertingly filthy when illuminated by a flashlight at night. Debris on the mirror will scatter light, and as the mirror gets grubbier it will be less effective, producing a decline in image contrast. Cleaning the surface of an aluminized mirror must be performed very carefully, since hard bits of debris scraped across the thinly aluminized surface will leave tracks like skates on ice. Loose debris can be blown away with a puffer or a canister of compressed air, and the mirror can be cleaned with cotton wool and lens cleaning fluid or lens wipes – this must be performed very gently, with a single stroke per cleaning wipe.

One way of extending the life cycle of a reflector is to stretch a piece of optically transparent film over the telescope aperture in order to seal the top of the tube (the bottom of a Newtonian is often open, allowing the free circulation of air for better image quality). This material is available in large sheets that can be cut to fit. For the best image quality, the sheet should ideally be taut, without wrinkles. When the film itself gets dirty, another disk can easily be produced. A well protected, well cared for Newtonian mirror can last more than a decade before requiring to be re aluminized.

A peek into Nos Ebrenn (Cornish: Night Sky), the author's 'observatory'. Inside are two 200 mm Schmidt-Cassegrains (an LX90 and vintage Dynamax 8) and a home-made 150 mm f/11 Newtonian, while outside is 'The Judge', a home-made 300 mm Newtonian (Credit: Grego)

11.6 Catadioptrics

Catadioptric telescopes use a combination of mirrors and lenses to collect and focus light. There are two popular forms of catadioptric – the Schmidt-Cassegrain telescope (SCT) and the Maksutov-Cassegrain telescope (MCT). SCTs are becoming increasingly popular. Light enters the top of the telescope tube through a large corrector plate – a sheet of flat-looking glass with a large secondary mirror mounted at its center. The corrector plate is actually aspheric in shape, figured to refract the light onto an internal primary mirror, which reflects light onto the convex secondary mirror, which in turn reflects light back down the tube and through a central hole in the primary mirror into the eyepiece. The relatively large size of the secondary mirror in a SCT produces a degree of diffraction that can slightly affect image contrast. An optically good, well-collimated SCT will deliver superb views of Mars. Moreover, due to their design, a number of useful accessories can be attached to the 'visual back' (the part of the telescope that the eyepiece normally fits into) – accessories useful to the Mars observer include filter wheels, SLR cameras, digicams, camcorders, webcams and CCD cameras.

MCTs use a spherical primary mirror and a deeply curved spherical lens (a meniscus) at the front of the tube. The secondary mirror in the MCT is a small spot aluminized directly on the interior surface of the meniscus. Light enters the tube through the meniscus, refracts onto the primary mirror and is reflected into the eyepiece via the secondary mirror and a central hole in the primary. Although MCTs superficially resemble SCTs, MCTs tend to be far better performers on the

Moon and planets. With their long focal lengths and excellent correction for spherical aberration, MCTs deliver excellent resolution, high contrast views of the Martian surface.

11.7 Telescopic Resolution

The bigger a telescope's objective lens or primary mirror, the finer the detail will be seen on Mars, but this is ultimately restricted by the quality of the seeing conditions (see below). On nights of really good seeing, the resolving power (R, in arcseconds) of a telescope of aperture diameter (D, in millimeters) can be calculated using the formula $R = 115/D$.

Aperture (mm)	Resolution (arcsec)	Suggested max magnification
30	3.8	60
40	2.9	80
50	2.3	100
60	1.9	120
80	1.4	160
100	1.2	200
150	0.8	300
200	0.6	400
250	0.5	500
300	0.4	600

1 arcsecond resolution — easily attainable by a 150 mm telescope — equates to around 280 km (the size of Lake Ontario) at the center of Mars' disk when the planet is at a perihelic opposition and around 25 arcseconds across

11.8 Eyepieces

A telescope can have the most perfectly figured lenses or mirrors, but it will not perform at its best if a poor quality eyepiece is used. An eyepiece's magnification can be calculated by dividing the telescope's focal length by the focal length of the eyepiece. A 20 mm eyepiece used on a telescope with a focal length of 1,500 mm will deliver a magnification of 75 (1,500/20 = 75). The same eyepiece will magnify ×40 on a telescope with a focal length of 800 mm (800/20 = 40).

It is recommended to have at least three good quality eyepieces that deliver low, medium and high magnifications. A low power eyepiece is excellent to view Mars in a celestial context, to view occasional close approaches to bright stars and deep sky objects. Using a telescope with a focal length of 1,000 mm a 20 mm focal length eyepiece with a 50° apparent field will show an actual field of around a degree and deliver a magnification of 50×. A high power eyepiece should deliver a magnification that is double the telescope's aperture in millimeters – for example, 200× on a 100 mm refractor. High powered scrutiny of Mars can only be performed when seeing conditions allow.

Spectacle wearers should consider the eye relief of any eyepiece they may want to purchase. Eye relief is the maximum distance from the eyepiece that the eye can comfortably be positioned to see the full field of view. Eyepieces with long eye relief allow the spectacle wearer to view in comfort without having to remove them in order to get the eyeball up close to the eye lens. Some eyepiece designs have better eye relief than others.

Eyepieces are produced in three barrel diameters – 0.965-in., 1.25-in. and 2-in. 0.965-in. eyepieces that come with many budget small telescopes are usually made of plastic and are of a very unsophisticated design and poor optical quality. Good 0.965-in. eyepieces are hard to find these days, so it's better to upgrade to 1.25-in. eyepieces. Most telescope focusers are built to accept 1.25-in. eyepieces, and some of them can accommodate 2-in. barrel eyepieces as well. 2-in. barrel eyepieces can be hefty beasts with incredibly large lenses. They usually accommodate very wide angle, long focal length optical systems that are ideal for deep sky observation.

Budget telescopes are usually supplied with Huygenian, Ramsden or Kellner type eyepieces which all have very restricted apparent fields of view – more a deep sea diving experience than a spacewalk! Huygenian, Ramsden and Kellner eyepieces are unsuitable for observing Mars at high magnification.

The Huygenian is a very old design, consisting of two plano-convex lenses; the convex sides both face the incoming light, and the focal plane lies between the two lenses. Huygenians are under-corrected (the rays from the outside zone of the lenses come to a shorter focus than rays focused by the central parts), but the aberrations from each lens effectively cancel each other out. Huygenians deliver very small apparent fields of 30° (or even smaller), and are only suitable for use with telescopes with a focal length of f/10 or greater. Huygenians have poor eye relief.

Ramsdens are another very old design, like Huygenians consisting of two plano-convex lenses, but both convex sides face each other (sometimes the lenses may be cemented together to provide better correction); the focal plane lies in front of the field lens (the lens that first intercepts the light). Ramsdens deliver flatter, slightly larger apparent fields of view than Huygenians, but they are prone to a greater degree of chromatic aberration, are poor performers on short focal length telescopes and offer poor eye relief.

Kellners are the least ancient of the three basic designs. Similar to the Ramsden, their eye lens (the lens closest the eye) consists of an achromatic doublet. Kellners deliver better contrast views than Huygenians or Kellners, with fields of about 40°, but annoying internal ghosting is invariably seen when viewing bright objects like Mars at opposition. Like the Ramsden, the Kellner's focal plane is located just in front of the field lens, so any minute particles of dust that happen to land on the field lens will be seen as dark silhouettes against Mars. Kellners have good eye relief, and those with focal lengths longer than 15 mm perform the best; shorter focal lengths can produce a blurred effect around the edge of the field along with chromatic aberration.

Monocentric eyepieces consist of a meniscus lens cemented to either side of a biconvex lens. Despite having narrow apparent fields of view of around 30°, monocentrics deliver excellent crisp, color-free, high contrast images of Mars, completely free from ghost images, and they can be used with low focal length telescopes.

Orthoscopic eyepieces comprise four elements – an achromatic doublet eye lens and a cemented triplet field lens. They have good eye relief, produce a flat, aberration-free field, and deliver excellent high contrast views of Mars. Their apparent field of view varies from around 30° to 50°.

Erfle eyepieces have multiple lenses (usually a set of two achromatic doublets and a single lens, or three achromatic doublets) that deliver a wide 70° field of view with good color correction. Erfles perform at their best when used with long focal length telescopes, and the best versions are of 25 mm focal length and greater. However, the definition at the edge of the field of view tends to suffer, and their multiple lenses produce internal reflections and annoying ghost images when bright objects like Mars are viewed.

Today most popular eyepiece is the Plössl, a four element design that produces good color correction and an apparent field of view around 50° that is flat and sharp up to the edge of the field. They can be used with telescopes of very short focal length. Standard Plössls with long focal lengths have a good degree of eye relief. Lower focal length Plössls of a standard design have poor eye relief, so they may be a little awkward to use for high magnification views of Mars, but versions are available with longer eye relief. If you are a little frustrated with having to refocus every time a different eyepiece is used, some ranges of eyepieces are parfocal, meaning that little or no adjustment to the focus is needed when you switch.

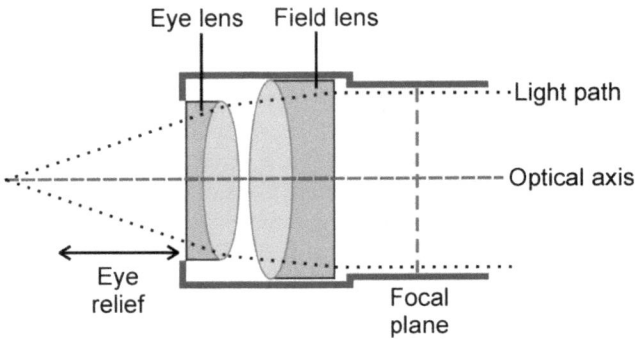

A simplified cross-section through a Plössl eyepiece (Credit: Grego)

Modern demand for quality wide field eyepieces has led to the development of designs such as the Meade UWAs, Celestron Axioms, Vixen Lanthanum Superwides and Tele-Vue Radians, Panoptics, Naglers and Ethoses, in addition to numerous clones of these eyepieces. They deliver excellently-corrected images with wonderfully large apparent fields of view, and all have good eye relief; short focal length versions can be used for medium to high power views of Mars.

Zoom eyepieces remove the necessity of changing between eyepieces of various focal lengths to vary the magnification. A number of reputable companies, including Tele-Vue, sell premium zoom eyepieces. Zoom eyepieces have been around for many years, but they have not yet achieved widespread popularity among serious amateur astronomers, perhaps because zooming is perceived to be a novelty associated with many budget binoculars and telescopes. Good quality zoom eyepieces are by no means joke items. They achieve a range of focal lengths by adjusting the distance between some of the lenses. A popular premium 8–24 mm focal length zoom eyepiece has a narrow 40° apparent field when set at its longest focal length of 24 mm, but as the focal length is reduced, the apparent field enlarges, up to 60° at 8 mm. A good zoom eyepiece can replace a number of regular eyepieces, and at a fraction of the cost. Mars can be zoomed in at leisure, but the telescope must be refocused each time the zoom is used.

A good range of eyepieces and accessories for the Mars observer. Clockwise from top: A 24 mm low power, wide angle eyepiece; an 18 mm wide angle medium power eyepiece; an f/6.3 focal reducer; a 9 mm high power Orthoscopic; a red filter; a 2× Barlow lens. With various configurations of these, no fewer than 18 different views of Mars is possible (Credit: Grego)

11.9 Binocular Viewers

Binocular viewers split the beam of light from the telescope's objective into two components which are reflected into two identical eyepieces. Most binocular viewers require a long light path, and they will only work on an instrument whose focuser can be racked in enough for the prime focus to pass through the convoluted optical system of the binocular viewer. A binocular viewer may not be able to focus through a standard Newtonian telescope, and the best instruments are refractors and catadioptrics (SCTs and MCTs). Binocular viewers are designed to be used with two identical eyepieces (at the very least, two eyepieces of the same focal length). Eyepieces of 25 mm focal length and shorter are recommended, since "vignetting" around the edge of the field becomes apparent when much longer focal length eyepieces are used. The use of two premium zoom eyepieces will save having to swap eyepieces to alter the magnification.

Viewing Mars up close with two eyes, rather than one, has distinct advantages. It is more comfortable to use both eyes, and the view is more aesthetically pleasing. With two eyes, a two-dimensional image takes on a near three-dimensional appearance, and many observers claim to be able to discern finer detail.

11.10 Telescope Mounts

It is important that a telescope is attached to a sturdy mount, and that the telescope is able to be moved without any difficulty in order to keep Mars in the eyepiece as the Earth rotates. The simplest form of mount is a telescope inserted into a large ball that can freely and smoothly rotate in a cradle – several small reflectors are mounted in such a manner, and they are great fun to use.

11.11 Altazimuth Mounts

Altazimuth mounts enable the telescope to be moved up and down (in altitude) and from side to side (in azimuth). Small un-driven table-top altazimuth mounts are often provided with small refractors, but the quality of their construction can be poor. Most problems are caused by inadequate bearings on the altitude and azimuth axes – they may be too small and the right amount of friction may be difficult to achieve. Overly tight bearings will result in too much force being used to overcome the friction, and smooth tracking cannot be achieved. Better altazimuth models are provided with slow motion knobs that allow the telescope to be moved without having to push the telescope tube around. If the mount itself is lightweight and shaky then it is liable to be buffeted by the slightest wind, rendering it unable to be used in the field – it may be better to attach the telescope to a good quality camera tripod.

Dobsonians are altazimuth mounts that are used almost exclusively for Newtonian reflectors of short focal length. Since their invention several decades ago, they have become highly popular because they are simple to build and easy to use. Dobsonians consist of a ground box with an azimuth bearing and another box that holds the telescope tube. The altitude bearing is at the center of balance of the telescope tube, and it slips neatly into a recess in the ground box. Low friction materials like polythene, Teflon, Formica and Ebony Star are used for the load-bearing surfaces, enabling the largest Dobsonians to be moved around at the touch of a fingertip. Lightweight structural materials such as MDF and plywood make Dobsonians strong but highly portable, and commercially produced Dobsonians range from 100 mm to half-meter aperture Newtonians.

It is not too difficult a task to keep Mars in the field of view of a telescope mounted on an undriven altazimuth or Dobsonian mount up to magnifications of 50×. The higher the magnification, the faster Mars appears to move across the field of view, and more frequent small adjustments need to be made to keep Mars centered in the field. If the observer wants to make an observational sketch of Martian features, the limit of magnification for an undriven telescope is 150× – anything higher and the instrument will need adjusting after each time the drawing is attended to – a tedious process that will double the length of time that a drawing should take to complete. At 150×, a feature that is centered in the field will take about 20 s to move to the edge of the field. High magnification with undriven telescopes also demands a sturdy mount that does not shake unduly when pushed, and smooth bearings that respond to light touch and produce little backlash – qualities found in only the best altazimuth and Dobsonian mountings.

The author with his 200 mm SCT (left) and home-made 300 mm Newtonian (Dobsonian mount) (Credit: Grego)

11.12 Equatorial Mounts

Serious Mars observation requires a telescope to be mounted on a sturdy platform with one axis parallel to the rotational axis of the Earth, the other axis at right angles to it. In an undriven equatorial telescope Mars can be centered in the field of view and kept there with either an occasional touch on the tube or the turn of a slow motion control knob that will alter the pointing of on one axis – far easier than having to adjust the telescope on both axes of an altazimuth mounted telescope to keep a celestial object within the field of view. A properly aligned, well-balanced driven equatorial allows the observer more time to enjoy Mars without worrying about it quickly drifting out of the field of view. Equatorial clock drives run at a 'sidereal' rate, enabling celestial objects that are centered in the field to remain there over extended periods of time, depending on how well the equatorial's polar axis is aligned, the accuracy of the drive rate and the apparent motion of the celestial object.

German equatorial mounts on aluminum tripods are most often used with medium to large refractors and reflectors. A telescope mounted on a German equatorial is able to be turned to any part of the sky, including the celestial pole. Schmidt-Cassegrain telescopes are commonly fixed to a heavy duty fork type equatorial mount. The telescope is slung between the arms of the fork, and the base is tilted to point to the celestial pole. When their visual back has a particularly large accessory attached, say a CCD camera, these instruments are sometimes unable to view a small region around the celestial pole because the telescope cannot swing fully between the fork and the base of the mount – this is no problem because Mars never gets anywhere near the celestial pole!

Many amateurs choose to keep their mount and telescope in a shed and set it up each time there is a clear night – setting up requires some time, and is usually done in several stages. The mount's polar axis must be at least roughly aligned with the celestial north pole for it to track with any degree of accuracy. A tripod can be difficult to adjust on sloping ground, and the legs not only pose something of a navigational hazard when walking around the instrument in the dark, but the seated observer will invariably knock the tripod from time to time, producing vibration of the image. To eliminate the time-consuming chore of setting up and polar aligning each observing session, some amateurs construct a permanent pier, set in concrete, upon which their German equatorial mount can be fixed and aligned, or to which their entire SCT and mount can be quickly and securely fixed.

11.13   Computerized Mounts

Computers are revolutionizing amateur astronomy in many ways, and one of the most visible signs of this is the increasing preponderance of computer controlled telescopes. They come in all varieties – small refractors mounted on computer-driven altazimuth mounts to large SCTs on computerized fork mounts and German equatorials. Some standard undriven equatorial mounts can be upgraded to accept either standard clock drives or computerized drives. When details of the observing location and the exact time are input, a computerized telescope can automatically slew to the position of any celestial object above the horizon at the touch of a few buttons on a keypad. Smaller computerized telescopes tend to have fairly insubstantial mounts that are incapable of supporting much more weight than the telescope itself and maintain a good pointing and tracking accuracy – while they are acceptable for visual Mars observation, they may not withstand the addition of a heavy accessory such as a digicam or binocular viewer. Larger computerized telescopes of the SCT varieties produced by Meade and Celestron, for example, are constructed well enough to accommodate hefty accessories. A computerized telescope can automatically slew to the position of Mars and track it accurately at the touch of a button, and basic Mars information can be displayed on the keypad's view screen. Unsurprisingly, many Mars observers with conventional equatorial mounts may not find these small advantages great enough to persuade them to make the upgrade. It is often argued that computerized telescope mounts are leading to a 'dumbing-down' of practical astronomy, since the convenience of being able to locate celestial objects eliminates the need for the amateur to learn his or her way around the skies and to star-hop to find the fainter deep sky objects. Such a debate will doubtless continue long into the future.

Resources

Societies

The Society for Popular Astronomy (SPA)

SPA website: http://www.popastro.com
   Address: The Secretary, 36 Fairway, Keyworth, Nottingham, NG12 5DU, United Kingdom.
   Email: membership@popastro.com
   Founded in 1953, the SPA is the largest astronomical society in the UK. It is aimed at amateur astronomers of all levels. The SPA produces an excellent bimonthly magazine, Popular Astronomy, which contains astronomy articles, news, notes and observing reports. The SPA hosts quarterly London meetings.
   SPA Planetary Section website: http://popastro.com/planetary/
   The SPA has a thriving Planetary Section directed by Alan Clitherow.

The British Astronomical Association (BAA)

BAA website: http://www.britastro.org
   Address: The Assistant Secretary, The British Astronomical Association, Burlington House, Piccadilly, London, W1J 0DU, United Kingdom.
   A UK based astronomical association aimed at amateurs with an advanced level of knowledge and expertise.
   BAA Mars Section website: http://www.britastro.org/mars/
   The BAA Mars Section is directed by Dr Richard McKim.

The Royal Astronomical Society (RAS)

Website: http://www.ras.org.uk
   Address: Royal Astronomical Society, Burlington House, Piccadilly, London, W1J 0BQ, United Kingdom.
   Founded in 1820, the RAS is the UK's leading professional body for astronomy & astrophysics, geophysics, solar and solar-terrestrial physics, and planetary

P. Grego, *Mars and How to Observe It*, Astronomers' Observing Guides,
DOI 10.1007/978-1-4614-2302-7, © Springer Science+Business Media New York 2012

sciences. Its bimonthly magazine, Astronomy and Geophysics, features occasional informative articles about the planets. Fellowship of the RAS is open to non-professionals.

Association of Lunar and Planetary Observers (ALPO)

Website http://www.lpl.arizona.edu/alpo
This large association, based in the United States, has a superb Mars Section and plenty of online resources and web links.

Unione Astrofili Italiani (UAI)

Website: http://www.uai.it/sez_lun/english.htm
Based in Italy, the UAI has active planetary observing sections, with a very informative English version of its website.

Internet Resources

Astronomy Picture of the Day: http://antwrp.gsfc.nasa.gov/apod/astropix.html
NASA/JPL Solar System Simulator: http://space.jpl.nasa.gov/
USGS Astrogeology Research Program, Gazetteer of Planetary Nomenclature: http://planetarynames.wr.usgs.gov/index.html

USGS Gazetteer of Planetary Nomenclature

http://planetarynames.wr.usgs.gov

NASA's Planetary Photojournal

http://photojournal.jpl.nasa.gov/targetFamily/Mars

Views of the Solar System

http://ftp.uniovi.es/solar/eng/homepage.htm

JPL Solar System Simulator

http://space.jpl.nasa.gov/

Books

The Planet Mars

By E.M. Antoniadi
  Publisher Reid (1975)
  Originally written in 1930, Antoniadi's work includes his own observations of Mars. Much of it remains very useful to visual observers today. It's now out of print, but if you can get hold of a copy you won't be disappointed!

Astronomical Cybersketching: Observational Drawing with PDAs and Tablet PCs

By Peter Grego
  Publisher: Springer (2009)
  ISBN-10: 0387853502
  ISBN-13: 978-0387853505
  A guide to electronic sketching of astronomical objects using handheld computers.

Solar System Observer's Guide

By Peter Grego
  Publisher: Collins (2006)
  ISBN-10: 0540088277
  ISBN-13: 978-0540088270
  An observational guide, contains a section on observing Mars.

The Compact NASA Atlas of the Solar System

By Ronald Greeley and Raymond M. Batson
  Publisher: Cambridge University Press (2002)
  ISBN-10: 052180633X
  ISBN-13: 978-0521806336
  A detailed reference work containing charts of Mars.

The Planet Mars: A History of Observation and Discovery

By William H. Sheehan
  Publisher: University of Arizona Press (1996)
  ISBN-10: 0816516405
  ISBN-13: 978-0816516407
  A beautifully researched history of Mars observation and a (now dated) outline of plans to explore the Red Planet.

Mars: The Lure of the Red Planet

By William H. Sheehan and Stephen James O'Meara
Publisher: Prometheus Books (2001)
ISBN-10: 157392900X
ISBN-13: 978-1573929004
A beautifully researched history dealing with Mars observers and their drive to understand the Red Planet.

Glossary

Albedo	A measure of an object's reflectivity. A pure white reflecting surface has an albedo of 1.0 (100 percent). A pitch black, non-reflecting surface has an albedo of 0.0.
Altitude	The angle of an object above the observer's horizon. An object on the horizon has an altitude of 0°, while at the zenith its altitude is 90°.
Aperture	The diameter of a telescope's objective lens or primary mirror.
Aphelion	The point in an object's orbit furthest from the Sun.
Apparition	The period of time during which a planet, asteroid or comet can be observed between conjunctions with the Sun.
Arcminute	One minute of arc. 1/60th of a degree. Indicated with the symbol'.
Arcsecond	One second of arc. 1/60th of an arcminute. Indicated with the symbol".
Asteroid	A minor planet. A large solid body of rock in orbit around the Sun.
Astronomical Unit	A convenient measure of distances within the Solar System, based on the average distance of the earth from the Sun. 1 AU is equal to 149,597,870 km.
Atmosphere	The mixture of gases surrounding a planet, satellite or star.
Axis	The imaginary line around which a planet rotates.
Basin	A very large circular structure, usually formed by impact and comprising multiple concentric rings.
Caldera	A sizeable depression in the summit of a volcano, caused by subsidence or explosion.
Catena (plural: Catenae)	A chain of craters.
Central peak	An elevation found at the centre of an impact crater, usually formed by crustal rebound after impact.

Conjunction	The apparent close approach of a planet to the Sun or another planet, seen from Earth. Mars is in conjunction with the Sun when the Sun lies between that planet and the Earth.
Crater	A circular feature, often depressed beneath its surroundings, bounded by a circular (or near-circular) wall. Almost all of the large craters visible in the Solar System have been formed by asteroidal impact, but a few smaller craters are endogenic, of volcanic origin.
Culmination	The passage of a celestial object across the observer's meridian, when it is at its highest above the horizon.
Dark side	The hemisphere of a solid body not experiencing direct sunlight.
Degree	As a measurement of an angle, one degree is 1/360th of a circle. Indicated by the symbol °. In terms of heat, degrees are increments of a temperature scale. Scales most commonly used in astronomy are Celsius (C) and Kelvin (K). 0°C, or 273.16°K, is the freezing point of water. 0°K, or -273.16°C, is known as absolute zero, where all molecular movement ceases.
Dome	A low, rounded elevation with shallow-angled sides. They can be formed volcanically or through sub-crustal pressure.
Eccentricity	A measure of how an object's orbit deviates from circular. A circular orbit has zero eccentricity. Eccentricity between 0 and 1 represents an elliptical orbit.
Ecliptic	The apparent path of the Sun on the celestial sphere during the year. The ecliptic is inclined by 23.5° to the celestial equator. Mars follows a paths close to the ecliptic.
Ejecta	A sheet of material thrown out from the site of a meteoroidal or asteroidal impact that lands on the surrounding terrain. Large impacts produce ejecta sheets composed of melted rock and larger solid fragments, in some cases producing bright ray systems.
Elongation	The apparent angular distance of an object from the Sun, measured between 0 to 180° east or west of the Sun.
Ephemeris	A table of numerical data or graphs that gives information about a celestial body in a date-ordered sequence, such as the rising and setting times of the Moon, the changing illumination of Mars, the longitude of Jupiter's central meridian, etc.
Equator	The great circle of a celestial body whose plane passes through its centre and lies perpendicular to its axis of rotation.
Fault	A crack in a planet's crust caused by tension, compression or sideways movement.

Gibbous	The phase of a spherical body between dichotomy (50% illuminated and Full (100% illuminated).
Graben	A valley bounded by two parallel faults, caused by crustal tension.
Highlands	Heavily cratered or mountainous region.
Impact crater	An explosive excavation in a planet's crust formed by a large projectile striking at high speed.
Lava	Molten rock extruded onto a planet's surface by a volcano.
Limb	The apparent edge of a planet.
Lithosphere	The solid crust of a planet.
Mare (Latin: Sea. Plural: Maria)	A large, dark albedo region.
Massif	A large mountainous elevation, usually a group of mountains.
Meteorite	A meteoroid which has survived its passage through a planet's atmosphere.
Meteoroid	A small solid body composed of rock or metal in orbit around the Sun.
Mons (Latin: Mountain. Plural: Montes)	The generic term for a mountain.
Occultation	The disappearance or reappearance of a star or planet behind the lunar limb.
Opposition	The position of a planet when its celestial longitude is 180° to that of the Sun.
Perihelion	The point in an object's orbit when it is nearest the Sun.
Phase	The degree to which a planet or satellite is illuminated by the Sun. Phases can be crescent (less than 50% illuminated) or gibbous (more than 50% illuminated).
Planet	One of eight large objects in orbit around the Sun, ranging from small solid Mercury to the large gas giant Jupiter.
Quadrature	The position of a planet when it has an elongation of 90° from the Sun.
Ray	A bright (though sometimes dark) streamer of material radiating from an impact crater.
Rift valley	A graben-type feature caused by crustal tension, faulting and horizontal slippage of the middle crustal block.
Rille	A narrow valley. Some rilles are linear, caused by crustal tension and faulting. Others are sinuous, believed to have been produced by fast-moving lava flows.
Satellite	A small body revolving around a larger body.
Secondary cratering	Craters produced by the impact of large pieces of solid debris thrown out by a large impact. Secondary craters often occur in distinct chains, where piles of material impacted simultaneously.

Glossary

Seeing	A measure of the quality and steadiness of an image seen through the telescope eyepiece. Seeing is affected by atmospheric turbulence, caused largely by thermal effects.
Solar System	The Sun and everything within its gravitational domain.
Sun	The central star of the Solar System.
Terminator	The line separating the illuminated and unilluminated hemispheres of a planet or satellite.
Universal Time (UT)	The standard measurement of time used by astronomers over the world. UT is the same as Greenwich Mean Time, and it differs from local time according to the observer's position on the Earth and the time conventions adopted in that country.
Volcano	An elevated feature built over time up by the eruption of molten lava and ash.
Zenith	The point in the sky directly above the observer.

Subject Index

Feature Index